I was there ... during that first second ... perhaps even before that.
Yes I was there ... in that quark soup ... and so were you.

TIME TO IMAGINE

AND OTHER SHORT 

SURENDRA KUMAR SAGAR

BlueRose ONE
Stories Matter
NewDelhi • London

BLUEROSE PUBLISHERS
India | U.K.

Copyright © Surendra Kumar Sagar 2023

All rights reserved by author. No part of this publication may be reproduced, stored in a retrieval system or transmitted in any form or by any means, electronic, mechanical, photocopying, recording or otherwise, without the prior permission of the author. Although every precaution has been taken to verify the accuracy of the information contained herein, the publisher assumes no responsibility for any errors or omissions. No liability is assumed for damages that may result from the use of information contained within.

BlueRose Publishers takes no responsibility for any damages, losses, or liabilities that may arise from the use or misuse of the information, products, or services provided in this publication.

For permissions requests or inquiries regarding this publication, please contact:

BLUEROSE PUBLISHERS
www.BlueRoseONE.com
info@bluerosepublishers.com
+91 8882 898 898
+4407342408967

ISBN: 978-93-5819-840-9

Cover design: Muskan Sachdeva
Typesetting: Rohit

First Edition: November 2023

CONTENTS

TIME TO IMAGINE

AN INTRODUCTION

ATMAN = BRAHMAN

`` *There Aren't Separate Selves That Inhabit A Single World* `` .. **Schrodinger**

JOURNEY FROM `TIME` TO `IMAGINE`

`TIME` is the name of the Apartment Complex in Indiranagar Bangalore where my wife Bharati and I lived for about 17.7 years from April 2004 to January 2022, `IMAGINE` is the name of the campus in Whitefield, Bangalore where our office (TOTAL ENVIRONMENT) is located.

Over the years, the average commuting time between TIME and IMAGINE was about 54 minutes in the mornings on the way to the office, and about 67 minutes in the evenings on the way back from office. The shortest time taken was about 12 minutes when my son Kamal was in the driver`s seat one morning in the year 2004 and the longest time taken was about 156 minutes one rainy day in the evening with plenty of traffic jams en route. But there was always great music playing in the car, so it was … well … enjoyable.

On one particularly bright and sunny day, while being driven from Residence (Time) To Office (Imagine), I wasn`t carrying a book or a newspaper that I always do, so to pass time I got myself involved in some calculations. I gave myself a problem:

``If all the matter in the universe is utilized to make a hotel, how many hotel rooms are there in that hotel? ``

This wasn`t at all difficult provided the assumptions are correct. I assumed the number of atoms in the universe as about 10^{81} (as per the scientist Arthur Eddington), the radius of an atom as 0.000000005 cm and the average room size as 4m x 4m x 3.2m, and the number of hotel rooms are worked out as $= (10^{81})* 1.33 * 3.14* (5*(10^{-9}))^3/ 4*4*3.2*(10^6) = $ About 10^{49} rooms.

Next question:

If the entire volume of all the space - including interstellar space as well as intergalactic space - in the universe is utilized as available space for constructing the hotel, how many hotel rooms are there in the hotel.

Apart from above stated assumption, we need to know the average density of matter in the universe. I am told it is about 0.2 atoms (excluding dark matter) per cubic. M of space (Martin Rees). So the empty space divided by space occupied by matter works out as
$=$

$1/ 0.2*1.33*3.14*(5*(10^{-9}))^3 = 10^{25}$

So the number of hotel rooms is worked out as $10^{49}*10^{25} = 10^{74}$ rooms

Next question : How many rooms are occupied by intelligent beings (couples)

Again we need to make some assumptions, here is a list of the assumptions:

Population of human beings (assumed intelligent) = about 8 billion.

Probable number of Intelligent civilizations who might have sometime lived in our galaxy `Milky way` as about 100000.

Probable number of intelligent civilizations currently living in the galaxy `Milky Way 'is about 1000 (The cosmologist Edward Harrison ... also see chapter five of Six Words), assuming also that the remaining 99000 intelligent civilizations have self-destructed themselves, as our own civilization is also planning to do.

Probable number of galaxies in the universe is about 100 billion (google)

Hence total number of couples occupying rooms = $8*10^9*10^3*10^2*10^9/2$

$= 4*10^{23}$ ...say 10^{24} (assuming my grandson does this calculation in the year 2100).

So how many rooms are occupied by couples?

Simple ...just one room out of every 10^{50} rooms.

What a waste of space

I reached K.R.Puram, still another about 25 minutes of travel time left.

What should I calculate?

Consider the trillions and trillions of cells in my body, and consider the quarks in each cell, or rather the couples, protons and neutrons (made up of three quarks each), that make up the nucleus of each atom.

And they all say : we all live in the same hotel, it has god knows how many rooms !

And that Hotel … Its me…. Hotel SKS … I am their universe.

Each cell is a hotel room … quite congested with atoms and molecules … but the couples in each atom are quite comfortable with considerable space to themselves …enough space for about 14000 trillion couples (assuming the radius of the nucleus as about 0.00000000000005 cm). ..Like a grain of sand living in a huge auditorium.

What a waste of space, not at all a cost-effective design.

On human being scales it works out just about three or four human beings living in the whole of Mumbai.

I`m thinking, that must be extremely lonely. How can we bring them together?

I guess you will have to go to a Neutron star and if you want to see something like that on the planet Earth, where all the electrons merge with their nuclei, the size of the Earth will reduce to that of a moderate size shopping mall with no reduction in its mass. Further if the Entire mass of the Earth is assumed as made up of human beings it will be something like Forty-five Billion Trillion men and women cramped within that shopping mall.

I`m thinking again, that will be very congested.

True, but not as congested as in a Black Hole where all of them will have to manage inside a Ping-Pong ball.

Traffic jam near Hoody Junction … but the music in the car is good … Benny Goodman

We move on, I think of these trillions of cells inside me, each one is a unit life in itself, each one is aware of itself in the singular, never in the plural, but I am the Hotel SKS where they all are staying inside me, and I am conscious of the fact that I am in possession of the sum total of the awareness of all these trillions of cells, but my own consciousness is somehow also in the singular.

I am thinking, just as I am conscious of the combined awareness of the trillions of cells inside me, there must be some sort of Super consciousness which is conscious of the fact that it is in possession of the sum total of the consciousness/awareness of the trillions of living creatures across the universe, and that that Super consciousness is itself in the singular.

It's just a question of scale.

And that Super consciousness is the Hotel where we all live.

And it's called `Hotel Brahman`

How the mind wavers.

I reach office, and as I walk to my room, I recall the equation:

ATMAN = BRAHMAN

That same evening, I described my day`s experience and wrote down my first Short Piece entitled ``ATMAN = BRAHMAN``.

From that day onwards, there have been many many days, when I, with books in the car and a calculator in my hand, when travelling To and fro from `TIME` to `IMAGINE` wasted no time at all and wrote or rather conceived these short pieces and then brought them to some shape during the ensuing weekends.

Most of the short pieces in this book were conceptualized during these travel times from `TIME TO IMAGINE`.

There is something for everyone and for all ages. The topics covered are wide ranging, on Science, Philosophy, Cosmology, Big Bang, Supernova, Electric life, Quantum Physics, Deep Philosophy, Teleology and Intelligent Field, Religious Philosophy, Inventory of Souls, Enlightenment in meditation, Geopolitics, The Doomsday Clock, Politics, American Presidential Election, Stock Market, Predicting the Sensex, Horse racing, Time Form Ratings, Waring limits in Pure mathematics, Randomness, Construction Engineering, Slip Forms, Family matters, Interesting Small World, The game of Scrabble, not forgetting Quiz Time and Jokes etc etc.

So, If I am asked the question: In which Genre I should consider this book to be, I will not be able to answer. I will leave it to the Publisher to decide.

Please don`t take these short pieces seriously, just have some fun.

But there is one short piece in the book which you might like to take seriously, and maybe do something about it. You will know which one as you read the book.

CHEERS AND BEST WISHES ... ...

A WALK IN THE GARDEN

``*So much of life, it seems to me, is determined by pure randomness.* `` —*Sidney Poitier*

An example, a hypothetical one, to illustrate how randomness makes it difficult to predict future events:

Rudolph was reading the book ``Critique of pure reason`` by Immanuel Kant. It was ten in the night. He decided to stop reading and go to sleep. As he went to close the main door he could feel the pleasant and invigorating breeze outside in the garden. Why not, he though, I should take a walk in the garden.

So, he took that walk in the garden, came back inside, closed the main door, bolted from inside, switched off the lamps, and went to sleep.

There was no way Rudolph could predict, that about one hundred and fifty years later, about twenty million people would die. All because of that walk in the garden. All because of a mosquito bite during that walk in the garden. All because of his falling sick. All because of his getting hospitalized as a result of his falling sick with malaria consequent on that mosquito bite during that walk in the garden. All because of the beautiful nurse that looked after him in that hospital.

Yes, indeed there was no way anyone could predict that twenty million people will lose their lives, all because of a man called Rudolph falling in love with a young and beautiful nurse who looked after him while he was under treatment in a hospital consequent on his falling sick with malaria as a result of a mosquito bite during the time he took a walk in his garden.

They got married, Rudolph and the nurse, and they lived happily thereafter.

They had a daughter, she was Hitler`s grandmother.

… … …

ARRIVAL

``I was there .. during that first second .. perhaps even before that.

Yes I was there .. in that Quark soup .. and so were you``

I entered the new aeon of the universe as a Quark....

What time was it?.

$10^{\wedge}$ (-32) seconds after Time zero .. I think

Something happened. Let's call it ..`THE BIG BANG`

`INFLATION` was `SWITCHED ON`

It all started at about 10^{-36} second after time zero, and during the period between 10^{-36} second and 10^{-35} second, the universe increased from a near-zero size to as much as a full centimetre across.

Was that considered high?

Unbelievably high, considering that even light, which travels at three hundred thousand kilometres per second, could not have travelled more than $3*10^{-25}$ millimetres in such a short time. The inflation lasted for a very, very short time, till the universe was just about 10^{-32} seconds old. Some scientists say it was even shorter. Opinions differ widely on this.

I believe it kept doubling itself every 10^{-35} fraction of a second, which means it doubled itself about a thousand times, even in this short duration of the inflationary epoch. In almost no time at all, it covered the entire observable universe, and in the process, it ironed out all the deformities in its way. In short, it wiped the slate absolutely clean and created all that space for the quarks to arrive on the scene. They did arrive in large numbers, perhaps 10^{82} of them, and these were the lucky ones who managed to escape annihilation by the antiquarks. The actual number of quarks and antiquarks was about thirty million times that number, and they annihilated each other all at once. However, one antiquark was short in each thirty million, with the result that about one out of every thirty million quarks could not find its antiquark and thus escaped annihilation. All the matter that is there in the universe (including you and me) is made up of just these surplus sets of quarks.

What next?

There was radiation all around. It was extremely hot, with a temperature about one hundred billion degrees kelvin. You entered the universe as a quark, and as I said earlier, you had but one chance in thirty million to survive. There were about thirty million quarks and the same number (minus one) of antiquarks in your neighbourhood who annihilated each other in no time at all. But one antiquark was short, and you failed to find your counterpart antiquark and thus avoided annihilation. So, each of us lost sixty million of our friends in no time at all. What could we do? It was programmed that way. If the antiquarks had been replaced by quarks, there would have been no annihilation and matter density of nearly sixty million times greater. The expansion would have stopped abruptly, causing instant crashing. The duration of inflation, depending upon all the initial conditions, was worked out precisely to enable the precise number of quarks to arrive on the scene, to obtain the precise density of matter per cubic meter of space per unit time elapsed since the beginning, and to obtain the precise rate at which the universe is expanding now, when it is about 13.7 billion years old. As the universe expands, the matter density reduces. As a rough guide, we use the rule of thumb that the average density of matter in the universe is inversely proportional to the square of its age. The current density of matter (actual matter) in the universe is about 0.2 atoms per cubic meter of space.

The question may be asked: Does that match well with the current expansion rate? For if matter was as scarce as that, the expansion would have been much faster. With the current rate of expansion, the matching matter content looks more like four atoms per cubic meter of space.

What then is the explanation?

I guess there must be something hidden from us that accounts for the remaining 3.8 atoms per cubic meter of space. something dark and hidden. Chances are that it is dark matter.

What happened to the quarks and antiquarks that annihilated each other?

That large-scale annihilation of matter/antimatter resulted in a stupendous burst of energy in accordance with the equation $E = mc^2$, and this is what brought about all this radiation. Our freedom as a single quark was quite short-lived; in less than a microsecond, two other quarks joined each of us, and together, as a trio of quarks, we got bigger and became a proton or maybe a neutron, with the former having an 80 percent probability and the latter about 20 percent.

What next?

So .. There was radiation all around .. It was extremely hot ..Temperature about 100 billion degrees Kelvin. By the time I was about one second old the temperature dropped to about ten billion degrees Kelvin and the density dropped to about 200,000 times the density of water, at this stage protons and neutrons started fusing into each other, there is a 25 % probability that a few of these joined me and I became an atomic nuclei, probably a Helium nuclei. This fusion process went on for about three and a half minutes and stopped abruptly as the temperature dropped to about one billion degrees Kelvin and the density dropped to about 20 times that of water. Most of the Helium that exists today was forged during this brief period. A substantial portion of the universe (about 75 %) did not participate in this fusion process and remained unprocessed which means there is a very good chance that I remained as a single proton Hydrogen nucleus.

Next came about 300,000 years of relative peace. We the Hydrogen/Helium nuclei along with free electrons and photons (Radiation) interacted constantly in thermal equilibrium, the temperature though reducing constantly was still extremely high, the light still strong and bright, the radiation still energetic enough to separate electrons from nuclei.

Expansion and cooling continued.

All of a sudden an abrupt change ..

The temperature had dropped to about 3000 degrees Kelvin. At this temperature the radiation was not strong or energetic enough to prevent free electrons from combining with nuclei and decoupled itself completely from matter and proceeded on its way through space at the speed of light losing temperature all the time. It has travelled for nearly 14 Billion years and still moving. It is there all around us and fills up all the intergalactic space. It is called the `Cosmic Background Radiation`, and the current temperature is just 2.73 degrees Kelvin. And so the electrons could no longer be prevented from combining with nuclei and becoming stable atoms. Accordingly, I became an atom. In the language of Quantum physics, you might say that I was in a state of superposition with an 80 % probability of being a Hydrogen atom and a 20 % probability of being a Helium atom.

And in the process, I got some mass with that "God particle" inside of me? The one they call "Higgs boson"? Actually, It was the media that called it the "God particle," and the scientists did not mind the publicity. In fact, it was neither God nor particle; it was a boson, and a boson is a force and not a particle. A proton consists of three quarks (two up quarks and one down quark) and of nothing else but these three quarks. The funny thing is that the proton mass is nearly hundred times the total mass of the three quarks, even though there is nothing else beside the three quarks in the proton. The three quarks add up to just about 1 percent of the total mass of the proton.

The question may be asked: Where does the remaining 99 percent of the mass come from?

Obviously, it comes from the strong force that binds the three quarks inside the proton. Try to pull them apart; try to separate them. You need an enormous amount of energy to do that. When you apply the equation $E = m * c ^ 2$, you know that there is mass over there. So, it's a boson and not a particle. Higgs was the scientist who thought of this first in 1964, and so they named it the Higgs boson.

So, the mass comes from the force, but where does the force come from?

Hmm I guess there is a 'field' out there in addition to the gravitational field and the electromagnetic field. Let's call it the Higgs field. In this Higgs field, trillions of quarks and antiquarks are arriving and then annihilating each other. In the end, nothing is left except the energy acquired from all the collisions, which becomes the force that binds the quarks that arrive (that is, those quarks that fail to find their antiquarks and thus do not get annihilated).

Such as those inside you and me?

Exactly. And when these quarks inside you and me arrived in this aeon of the universe, during that first microsecond after the big bang, all they did was just progress through that Higgs field.

So, the mass is actually nothing other than a manifestation of fundamental particles

trying to progress through the Higgs field.

Now this radiation as well as matter had ample time to communicate with their counterparts in different regions of the universe prior to the decoupling stage, and this aspect along with the fact that inflation earlier on had resulted in a stupendous zone of `Causal Contact` accounts for the fact that the Universe is homogenous all around and every entity be it Photons(Radiation) or atoms knew exactly how they were supposed to act or behave in the future and could not act or behave in any different way as there were no such inputs supplied to them to enable them to act differently.

Thus at age 300,000 years, I an atom along with others entered the next phase of my life knowing precisely what I had to do with the laws of Physics spelt out to me in no uncertain terms by the Anthropic Principle which in turn was designed and programmed by the cosmic consciousness of the previous universe.

And the first thing I did along with friends in my neighbourhood (while still moving with the expanding universe) was to collapse into something bigger, which in turn collapsed into something even bigger (heavier may be a more appropriate term), which in turn collapsed into something even bigger, and so on.

Call it `Gravitation` if you like ..And it went on and on for millions and millions of years .. .is still going on

....

THAT .. MY DEAR FRIENDS .. WAS A SUPERNOVA

``There's one tiny little gap in the universe left, just about to close. And it takes a lot of power to send this projection. I'm in orbit around a supernova. I'm burning up a sun just to say goodbye.``

Russell T Davies

`*Soon there will be no fuel left .. It will be time to escape .. but I must change my uniform ..change into carbon*`

So, when I was about 3.5 billion years old, I found myself to be a member of a certain Star in a certain galaxy called `**Milky way**`. It was a massive star, nearly 25 times as massive as the `Sun`. We (me and my colleagues) had been living there for nearly 25 million years while burning ourselves one by one from Hydrogen to Helium, and about to witness the last days of the star leading to an event of stupendously catastrophic magnitude and mind-blowing significance in terms of the next stage of the Anthropic Principle. We had reached a stage when the entire central core was converted to Helium. My own status at the time would have been either of the following:

a) I was already a Helium atom near the centre.

b) I got converted from Hydrogen atom to Helium atom.

c) I was still a Hydrogen atom being part of the remnant Hydrogen still unprocessed near the surface.

Then came the next stage being the initiation of the process leading towards the catastrophic death of the star.

What exactly happened in the next stage?

IGNITION OF THE HELIUM

Balance life of the star : About one million years.

Sharp rise in temperature in the already hot star .. Sharp increase in the density .. Contraction .. Conversion into Carbon and Oxygen.

This goes on till .. Temperature reaches about a Billion degrees .. Density reaches about 100,000 Tons/Cubic .Metre .. Most of the Helium in the central core is converted.

What's my status? .. Either an Oxygen atom in the central core .. Or a Carbon atom in the region surrounding the central oxygen core .. Or a Helium atom left unprocessed in the region surrounding the Carbon region .. Or a Hydrogen atom still unprocessed near the surface of the star.

What next /

BURNING OF CARBON

Balance life of the star: About 10000 years......

Contraction stops temporarily .. Temperature stops rising temporarily .. Carbon burning starts .. Conversion into magnesium.

This goes on for nearly 10000 years .. When most of the Carbon is burnt out .. Nucleosynthesis picks up speed.

Death is now very close

What next?

FORMATION OF IRON

Balance life of star: .. Few days

Contraction resumes at a highly accelerated rate .. Temperature starts rising again stupendously .. Nucleosynthesis proceeds at a staggering rate.

Formation of heavier elements like Silicon, Sulpher, Nickel, Cobalt, IRON, near the center.

Star takes the shape of an Onion with several layers of progressively heavier elements towards the centre with ignition taking place at the base of each layer and converting the elements in the layer to heavier elements in the layer below.

Central core dominated by **IRON.**

Balance life of the star: .. Two minutes

What's my status? .. An atom of IRON with maximum probability, or of Silicon, or Sulpher,or Oxygen, or Carbon, or Helium, or Hydrogen etc, with progressively reduced probabilities.

What next?

Stupendous struggle against gravitation collapse .. Creation of Neutrinos in Phenomenally huge numbers (Probably of the order of $10^{\wedge}57$) .. Escape of Neutrinos through the star into space leaking energy into space thus not permitting energy to support the star against collapse.

What next?

THE ULTIMATE COLLAPSE

Massive IRON core formed. Considering IRON as the most energetically favored element, its fusion into still heavier elements such as Gold does not release energy, but absorbs energy. as such it cannot squeeze energy from fusion and so there is nothing to prevent its gravitational collapse. In no time density increases to 10 billion Tons/Cub.M

Balance life of star: .. ONE SECOND

In less than a second density near the centre increased tenfold to about 100 Billion Tons/Cub.m.. That's about 50000 Billion human beings cramped in a normal size room.

I couldn't bear it anymore. I guess it was too damn overcrowded.

We had to revolt and we did just that, we revolted and crashed out. What else was there to do?

It was an EXPLOSION, a SUPERSHOCK of a million mind boggling magnitude.

Consider the equation $E = M*c^2$ with M as equal to several solar masses, and `c` of course as the speed of light, .then try to imagine the quantum of energy created in the explosion.

THAT MY DEAR FIENDS WAS A ..

SUPERNOVA

What has been captured in the above description of a typical Supernova is just a broad outline. The phenomena itself is too wide ranging in its character and scope depending among other things, on the star's mass (in terms of number of solar masses) with varying end results such as the star losing most of its material into space but retaining some of it in the heavily contracted central mass and becoming a Neutron star or even a Black hole (a victory for gravity) or disappearing altogether, losing all its material by expulsion into space(a victory for entropy).

Was this a performance to empty stalls?

No, it wasn`t, It would be watched by intelligent beings of the future in far off worlds.

Was this an industry?

Yes, it was, an industry to manufacture the elements destined to become the raw materials for future rocks, trees, buildings, tables, chairs, books, birds and animals, you and me. Nowhere else you'll find them except as products of Supernova.

It was the next stage of the `Anthropic Principle`

And I became an atom of Carbon.

… … …

ELECTRIC LIFE

` The soul is like electricity .. It's a force that can light a room `.. Ray Charles

AND SO, I BECAME AN ATOM OF CARBON

Subjected to the stupendous shock, I was at first dismembered to my original state (as in my birth second) of a Proton made up of three Quarks, and later joined by other friends, Protons, Neutrons, Electrons with their own but somewhat similar past history, and proceeded onwards on my journey, with absolutely no idea where I was headed to

Suddenly I noticed something ..

I felt different ..

What was it?

At the precise moment - in the aftermath of the Supernova - that I became an atom of Carbon a certain quantum of awareness crept in and became an integral part of me being the information content of about 3.5 Billion years of experience (culminating in the Supernova). But it was just an awareness, of a somewhat different kind. Coupled with this awareness was a property comprising of a special characteristic being the desire/and tendency to grow and BOND with other elements and this desire will someday get fulfilled .. God knows when .. I didn't know at the time the extent to which the odds were stacked against me …that I must wait millions (or perhaps billions) of years before I attract and bond with other elements and become .. `something` .. comprising of C….H…..O…..N…and other friends .. in a precise mix…

under a precise environment .. .at a precise temperature .. as fixed by the designers. Once formed this .. `something` .. would have the following characteristic features: It would be indestructible ..Its awareness quantum would increase by leaps and bounds over that of Carbon alone .. and it would be supremely intelligent .. as per the designed program .. but not conscious in the same way as in living beings… It would forever be on the lookout for a specific environment with a specific biochemistry to obtain a different kind of consciousness. And so, at time `t` = about 10.5 billion years after time zero I arrived on the planet Earth in search of that biochemistry.

When I landed on the planet Earth perhaps about three billion years back, it was still boiling. Some of my companions fell deep beneath the surface, but I remained closer to the top. I was sometimes part of a landmass, sometimes part of an ocean sea floor, sometimes part of a swamp. And this continued for millions and millions of years. Most of my colleagues in my neighborhood remained inert. But something peculiar happened to me, I could sense that an electric field was leaking out from me, from the charged electrons and nuclei inside me. The stretching electric fields from the electrons inside me, made me writhe and twist and pull myself into strange sort of configurations. This strange phenomenon happened to many of my colleagues in my neighborhood but most of them fell apart… But not me … I was made of sterner stuff … I was determined not to let go this golden opportunity. I had waited more than ten billion years for this, and there was no way I was going to miss out on this… I looked up at the Sun, absorbed the heat energy, and allowed the electric charges to take control over me and further contort the atoms inside me and allow myself to writhe and twist, and pull myself to another configuration; similar to the first … In short I had duplicated myself …. And then quadruplicated … and so on` `.

The rest is history….

And so, I take Another forward leap, and come to my present state of consciousness. At present I am residing in the Cerebral cortex of the Cerebrum or may be in the Reticular Formation or Hippocampus or Medulla (whatever) of the brain of a man called `SKS` dictating this resume and directing the brain to coordinate the flow of information (whatever), and causing the fingers of his hand to type on the laptop. Looks as if… I travelled for nearly 14 Billion years covering nearly ten billion trillion miles before realizing I travelled so much.

It has taken that much time (14 billion years or so) to reach this state of consciousness where I've had the privilege to develop an interest in Cosmology, to be alive at the time when this branch of Science has progressed by leaps and bounds, and to have interacted (through books of course) with the great Scientists and Cosmologists of the past and present. Now it's another matter that I may not have been consciously aware of my existence all these 14 Billion years and it's quite possible and even likely that there might have been long gaps of perhaps millions of years between successive consciousness states, even so it can be said that I have been forever conscious for the simple reason that I was unconscious of my unconscious tenure which period must have passed instantaneously for me despite being millions of years.

With all this .. `I AM ALL IN ALL` The world extended in space and time is just my representation. Experience does not give me the slightest clue of it being anything besides that. The rest of the universe is just a stage created for me by my own consciousness … It is just my representation.

And I am forever conscious, I cannot recall asking the question .. Where is my universe? `.

Yes indeed I am forever conscious, forever alive and kicking, and I will never ask the question: What comes after death? For, I know the answer. Its not the first time I am going to die … whenever that happens. .I have died many times before .. But I am living now … and it's a wonderful life … and it has turned out to be `what comes after my last death`

Perhaps, in the intervening period, the process of recharging goes on .. with the kind help of the `Electrons in the Intelligent Field`.

In short .. this might be an `**ELECTRIC LIFE**`

… … …

THOSE WILL BE THE DAYS

And so, I opened at random a page from the newly acquired book ``Critique of Pure Reason`` by Immanuel Kant, and read a para from the page ...

This was it:

`` *Now, since the notions of good and evil, as consequences of the a priori determination of the will, imply also a pure practical principle, and therefore a causality of pure reason; hence they do not originally refer to objects (so as to be, for instance, special modes of the synthetic unity of manifold of given intuitions of one consciousness) like the pure concepts of the understanding or categories of reason in its theoretic employment ; on the contrary they presuppose that objects are given; but they are all modes (modi) of a single category, namely that of causality, the determining principle of which consists in the rational conception of a law which as a law of freedom, reason gives to itself, thereby a priori proving itself practical. However, as the actions on the one side come under a law, which is not a physical law, but a law of freedom, and consequently belong to the conduct of beings in the world of intelligence, yet on the other side as events in the world of sense they belong to phenomena; hence the determinations of a practical reason are only possible in reference to the latter, and, therefore in accordance with the categories of the understanding ; not indeed with a view to any theoretic employment of it, i.e, so as to bring the manifold of (sensible) intuition under one consciousness a priori...But only to subject the manifold of desires to the unity of consciousness of a practical reason, giving it commands in the moral law, ie , to a pure will a priori* ` `

I am sure you`ll understand my reaction …

I opened a few more pages at random … the same reaction …

And then I asked myself the question: `Why did I buy this book?`

But then I had asked for it, so how could I refuse to buy it?

But why did I ask for it ? That is the question!

So here is the story:

It was a bright and sunny October Sunday afternoon; the year was 2012. We (with wife and son) had lunch at a restaurant somewhere at Lincoln Square in Chicago, and then decided to take a casual stroll on the street, looking around at the beautiful shops.

We entered a bookstore, I remember the name `Raven`s woods` … All kinds of books ……awesome variety of books ….It was`nt a big shop, but the concentration of books per square metre of the floor area was the highest I had seen so far. Being a structural engineer, I did some mental calculation and decided if ever I was to design a building floor for a bookstore it will have to cater to a live load of 700 kg per square meter, about 40 percent higher than that prescribed by the codes.

Anyway, we spent a good 30 to 40 minutes leisurely going through the books, and then the owner came forward and asked the question : `Are you looking for something in particular ?`… I was`nt actually… so I asked him (thinking there was less than 1% percent chance that he would have it) :

Do you have `The Critique of Pure Reason....by Immanuel Kant` ?

I was astonished when he said : `Yes I have it ......you won`t find it ...I`ll help you ..`.

Even he took a few minutes to locate it....It was bound and in good condition....printed in 1955...it was the second edition. Written in 1787 (first edition written in 1781) ... Price $8.00.. I tell you...It was a steal.

I had heard it was a famous book on philosophy, and that it was a difficult book to understand and it was possible that no one has read the book in completion...or maybe it is a single digit number. It is learnt that a person who read about half the book ...and understood it well ....presumably after a number of readings....once said : ` Whosoever has not read the book is still a child`

When asked why he did not complete the book...he promptly replied : `I would have surely become insane if I had done so`

Anyway, I decided to postpone the reading of the book for some time. Those days I was deeply involved in the understanding of Quantum Physics and Relativity, trying to understand things like `Non objectivity`,... `The observer created reality`...`Particle wave duality` ... `The collapse of the wave function` `The Probability wave`, `Quantum entities at two places at the same time`....most of all, `The non local action` `Quantum entanglement ....etc etc` .. That was enough complexity to deal with, and there was no point taxing the brain any more than its limited capacity.

And so I put it on the shelf and let it be there ..

But only for a day !

The next day, I was drawn to it .. in particular the three words `Unity of Consciousness` from that para.

I read the para once again … and again … and again with some concentration. Believe it or not, I was beginning to understand it.

And then I read it again with some more concentration, and `Bion`….it was nothing short of `Spellbinding`.

This is what I thought was the gist of what it meant :

`Before we make our judgment of whether something is `good` or `evil`, we must ask ourselves under which law we are making the judgement, whether it is the law of freedom (of will)…or it is a physical law ( law of Physics). The former from considerations of `Pure Practical Reason` corresponds to the conduct of beings in the `world of intelligence`, and the latter from considerations of Pure Reason (Theoretical) would mean that any act performed is in the world of sense..ie just a phenomena ie. merely the motion of atoms and molecules in accordance with the laws of causation.`

And when I think, there are about 1800 such paras in the book to be understood and then simplified, the realization manifests itself that `Life is too short`

So I look forward to the future … when the mind is not so full, when all the structures are completed, when all the books have been written, when I can lead a relaxed life with my better half .. with nothing to do but listen to music, take morning and evening walks, eat good food, and read ``The Critique Of Pure Reason``

Those will be the days …

… … …

`OPPENHEIMER` AND THE CORELATION WITH `BHAGWAD GITA`

OPPENHEIMER is considered as the father of The Atomic Bomb, but I think that Albert Einstein should be called `father` or maybe `Grandfather` of the AB.

Einstein`s equation `E= m*c^2, which implied that something even of a very small mass had the potential of creating a huge amount of energy, was of enormously far-reaching significance. The ball was set to roll for experimental research on the subject. Scientists and Science Fiction writers started imagining the consequences of utilizing the energy that gets release in radioactivity. H.G. Wells .. the author of the famous ``TIME MACHINE`` wrote a book entitled ``THE WORLD SET FREE`` where the words `atomic bombs` were used for the first time. Politicians and World leaders started predicting the future war scenarios.

The scientist Leo Szilard was among the first to realize its implication. While crossing a certain street in London – Where Southampton Row passes Russell Square, across from the British Museum in Bloomsbury – in Szilard`s words ``As the Light changed to green and I crossed the street, it …. Suddenly occurred to me that if we could find an element which id split by neutrons and which would emit two neutrons when it absorbs one neutron, such an element, if assembled in sufficiently large mass, could sustain a nuclear chain reaction. I didn`t see at the moment just how one would go about finding such an element, or what experiments would be needed, but the idea never left me. In certain circumstances, it might be possible to set up a nuclear chain reaction, liberate energy on an industrial scale, and construct atomic bombs``.

But the real credit for the Discovery of FISSION - During a certain walk - goes to the Scientist `LISE MEITNER`.

A certain walk in the snow-covered woods outside Stockholm was to shape the future of the human race. It was a walk during which the walkers, scientists Lise Meitner and her nephew Otto Frisch, discussed the findings of experimental research carried out by another scientist, Otto Hahn. During this discussion, they discerned something of extreme significance related to the splitting of the uranium nucleus and its consequent release of energy.

In that respect ``Lise Meitner`` can be considered as the `Mother` of the Atomic Bomb.

And then again It was Einstein who wrote letters to President Roosevelt in August of 1939 and March of 1940, warning of the German research in fission and urging the president to initiate and then speed up a research program in the United States to explore the feasibility of atomic bombs. These letters were considered by many as the biggest and most powerful interactions between Einstein and the United States. The German occupation of Czechoslovakia had brought a halt to the sale of uranium, and evidence pointed to intensive nuclear research by German scientists. The two letters, along with a third letter that resulted in the start of the Manhattan Project, contributed significantly toward enormous atomic research effort in the United States that ultimately tilted the scales in favor of the United States in the race toward the first use of the atomic bomb.

However, it goes without saying that Oppenheimer, an exceptional theoretical physicist as well as General Leslie Groves chosen to head the Manhattan Project played a stupendous role, and the two were indispensable to the success of the Manhattan Project.

Now coming to the Corelation with Bhagavat Gita, Oppenheimer had a long and documented fascination with the *Bhagavad Gita*. Oppenheimer also had a "taste for the mystical, the cryptic"—things like the ancient Hindu text, a dialogue between a human hero named Prince Arjuna and Krishna, Reincarnation of the Lord Vishnu. When Oppenheimer first read the text, he couldn't have known the eerie relevance it would have to his future. "About to lead his troops into mortal combat," "Arjuna refuses to engage in a war against friends and relatives. Lord Krishna replies, in essence, that Arjuna must fulfil his destiny as a warrior to fight and kill."

When however, it is a question of justification for war, we simply cannot compare the two cases. Oppenheimer had absolutely no choice whatsoever. The decision to drop the bomb was not his at all, it is true and it was quite obvious that the world would not be the same after the dropping of the bombs, but it cannot be denied that the other alternative of a German First use of the Bomb would have been far more catastrophic for the world.

On the other hand, Lord Krishna`s advice to Arjuna and trying to persuade the prince that he should do his duty as a warrior, tantamount to taking sides is questionable. Lord Krishna`s contention was that Arjuna is a soldier; he has a duty to fight. Krishna, not Arjuna, will determine who lives and who dies, and Arjuna should neither mourn nor rejoice over what fate has in store, but should be sublimely unattached to such results.

In the eyes of Krishna, the Godhead, or `Isvara` the supreme controller, both Duryodhana and Arjuna are normal human beings `jivas` ie the controlled living entities. They were the controlled ones; they were not the controllers nothing was in their control.

Under the circumstances why should Krishna have differentiated and taken the Kaurva`s as wrongdoers and the Pandavas as all Pious gentlemen. Besides the question may be asked: What exactly was so Pious about the Pandavas. They were all gamblers to the core, what's more, they put their wife at stake. This is simply not acceptable to the thinking mind. There is nothing pious about it.

Besides, the question may be asked: Who needs Krishna`s advice more ? Duryodhana or Arjuna? If Arjuna was the sensible one and knew what is right and what is wrong, whereas Duryodhana had no understanding of what is right and what is wrong, then he was the one who needed to be advised. As the SUPREME CONTROLLER, Krishna should have persuaded both parties to come together in a conference and resolve their disputes, rather than go to war and allow tens of thousands of innocent soldiers of the two armies to get killed for no fault of theirs at all.

On the topic of Isvara it is my considered view that Isvara is not an anthropomorphic physical function God. An anthropomorphic God, made up of atoms and molecules, occupying time and space, a miracle performing, rewarding and punishing God is not compatible with the established principles of Science. . A God who rewards and punishes is inconceivable for the simple reason that a man`s actions are determined by necessity, external and internal, so that in God`s eyes he cannot be responsible, any more than an inanimate object is responsible for the motions it undergoes.

Equally impossible to believe is the concept of `No God at all`, as that would mean that nature acts on its own. This would lead to a universe that makes no sense at all. The only other alternative is that there is `Something` out there which controls nature in such a way as to ensure that the Universe makes sense. It is not known what is that `Something` that controls nature. It may take a long, long time for Science to understand what this `Something` is. No harm will be done if we call this `Something` as God. And no harm will be done if we assume the nature of this God as an `intelligent field` However, the scope of work of the `Intelligent Field` (IF) is limited to creating consciousness only. There is nothing `divine` about `IF`. It acts according to the `Information in the Field`. It is the Information in the field which is responsible for the phenomenon.

RELIGION AND WAR

Coming to the conflicts between nations which are predominantly Religious based, in my view, it is very important to develop a better understanding of the requirement of a universal religion, which will have no location in place or time, which will be infinite like the God it will preach. It will be a religion which will have no place for persecution or intolerance in its polity, which will recognize divinity in every man and woman, and whose whole force will be created in aiding humanity to realize its own true divine nature. The power of religion, broadened and purified, is going to penetrate every part of human life. Therefore, religions will have to broaden. Religious ideas will have to become universal, vast and infinite.

Adopting a Universal Religion is the prime requirement, and even if there are differences of opinions about the nature of God, or the form in which God exists, all religions must converge into a universal religion that teaches us that God is a mind. I believe that `One ness of the mind ` or the `Unity of consciousness` is the central idea of universal religion.

Each individual on the planet must have the freedom to choose his or her own religion and that freedom should include freedom from religion.

Back to Oppenheimer and the nine words of the Bhagwat Gita : 'Now I am become Death, the destroyer of worlds.' And Oppenheimer`s words: ``I suppose we all thought that one way or another``, This aspect may be viewed in the light of the fourth argument of the Bhagwat Gita.

"The fourth argument in the *Gita* is really that death is an illusion, that we're not born, and we don't die. That's the philosophy, really: that there's only one consciousness, and that the whole of creation is a wonderful play`` "I am become Death," then, is not necessarily a quote about destruction.

I am in total agreement with the philosophy of the ``fourth argument`` which says that there is only one consciousness. Indeed `Oneness of The Mind` , in other words ``unity of Consciousness`` is the central theme of my two books ``SIX WORDS`` and ``INTELLIGENT FIELD``, which in my view should be the central idea of a Universal Religion. There is nothing Hindu, Muslim , Christian, about it just as there is nothing Greek about the Pythagoras Theorem.

… … …

Sometime in the Eighteenth century, there lived a man named Bhoj Raj Saggar in the Indian State of Punjab.

So! What`s the big deal ?

Will come back and explain.

… … …

In the meantime, here is another story:

This part is imaginary.

The year was 2017. Two men in their mid-eighties, a gentleman called `CKM` and another called `BS` find themselves sitting next to each other watching `GOO` (Gran Ole Opry) at Nashville US.

They are strangers, but soon start talking with each other in appreciation of the music. The last number on the programme: `Rock Island Line `

CKM: That one is a `skiffle`

BS: What is a skiffle?

CKM: Where the singer is sometimes talking and sometimes singing, and the beat picks up gradually, and usually there is some sort of a story line.

BS: Was `Beep Beep`.. by the Playmates a skiffle?

CKM: Yes, that's right.

They are enjoying the skiffle sung and enacted by the singer ..

It goes something like this :

` `Now this here`s the story `bout the rock island line .. The rock Island line she runs down to New Orleans .. And just outside New Orleans .. There`s a big toll gate .. And all the trains running up to the toll gate .. Why ! they gotta pay the man some money .. But of course if you have certain things on board..You`re ok, you don`t have to pay the man nothing.. And right now we see a train coming down the line.. And when the train she nears the toll gate.. The driver, he shouts to the man. He says : ` ` I got horse, I got sheep, , I got cows, .I gotta whole life stock, I gotta whole livestock, I gotta whole live stock ..

So you can hear the man say: `You`re alright boy .. You just get through that door.. and you don`t have to pay me nothing.

And so the train moves through.. And as the train moves through .. She takes up … a little bit of steam .. And a little bit of speed ..And when the driver.. He thinks he is safely on the other side .. He shouts back to the man .. he says :`I fooled you .. I fooled you .. I fooled you … I got Pig ions.. I got pig ions .. I got aaall pig ions` ..

Music ..

Now this is where I `m going boy ..

Music ..

`The Rock Island Line is a mighty good road .. Rock Island Line is a road to ride, Rock Island Line, well it's a mighty good road .. Well if you want ..You wanna get a riding like you find it get your ticket at the station for the rock island line ` … (a)

Music …. Repeat `a` ..

A B C to the X Y Z …The cats in the cupboard but you can`t see me.

Repeat `a` ..

Music …` ` `p p p pau pau pau ..pppp a pau pau pau, p p p pau pau pau ..pppp a pau pau pau, p p p pau pau pau ..pppp a pau pau pau,

Pappa puppa pau, Pappa papaa pappa pappa pappa ..papapuppup` `.. (b) .. (only in my version)

Repeat `a`

Now I may be right .. or I may be wrong ..

But you`re gonna miss me when I am gone .. Ohh

Repeat `a`

Music ..

Repeat (b) … the beat getting faster

Repeat `a`

Music …

Repeat `b` …Faster and faster

Repeat `a` … real fast` `

The music ends..

Applause … continues … and becomes a standing one

CKM: Wasn`t that great ?

BS : Yes .. was`nt that a Johnny Cash number?

CKM : But the original version was by Lonnie Donnagan .. I think

The programme ends ..

They all shake hands ... Strangers before the show... friends after the show ...

The bye bye `s

They exchange visiting cards.. phone numbers .. websites .. and so on.

BS looks at CKM`s visiting card in astonishment, and here is what he says:

``Oh my God ... you are CKMeador ... I heard so much about you ... And I know who you are ... and we are relatives.. doesn`t matter how distant .. This indeed is a small world .. not just in space but also in time!``

CKM couldn`t understand and asked the obvious question : I don`t get it .. who are you ?

BS : First I`ll tell you who you are .. and then you can work out who I am

CKM: Go on

BS : My dear Dr.CKM You are my Dad`s Dad`s Dad`s brother`s son`s son`s son`s son`s wife`s dad`.

CKM : which means you are my daughter`s husband`s, dad`s dad`s dad`s dad`s brother`s son`s son`s son.

This is amazing indeed.

... ...

Wasn`t that interesting ?

Now, I come back to this man called Bhoj Raj Saggar who lived in the Eighteenth century in the state of Punjab, India.

And this is a true story ... all records available in a place called Hardwar in North India.

Punjab is a state in Northern India ..Americans and Europeans may have for the first time heard about Punjab in 1958 ... remember the song ``Boomboodi, Boomboodi, Boomboodi, Boomboodi Babam bam ... Goodness Gracious Me .. a duet by Sophia Loren and Peter Sellers``

Now this man Bhoj Raj Saggar had a son, his name, Dhanpat Raj Saggar.

And one of Dhanpat Raj Saggar`s sons was a man called Balak Ram Saggar.

And one of Balak Ram Saggar`s sons was a man called Sukh Mal Saggar.

And one of Sukh Mal Saggar`s sons was a man called Hira Lal Saggar.

Now this man Hira Lal Saggar ... I don`t know how many sons or daughters he had, but three of his sons were named Shriram Saggar, Nanak Chand Saggar, and Munshi Ram Saggar.

Will come back to talk about Shri Ram Saggar, but here is something about Nanak Chand Saggar:

Now this man Nanak Chand Saggar, I don`t know how many sons or daughters he had, but one of his sons was a man called Kishori Lal Saggar.

This gentleman called Kishori Lal Saggar, decided to remove one g, and so from this point onwards the Surname Saggar became Sagar.

Kishori Lal Sagar, an Engineer from Thomson College Roorkee (Currrently I.I.T Roorkee) had nine sons and their names in descending order of Age are ``Pyare Lal Sagar, Atma Ram Sagar, Jamna Das Sagar, Raj Roop Sagar, Sohan Lal Sagar, Om Prakash Sagar, Rawel Chand Sagar, Karam Chand Sagar, and Gyan Chand Sagar.

All those whose names have been mentioned so far .. are no more .. and their souls are either resting in peace .. or trapped in other human bodies somewhere on the planet Earth.

Of the nine brothers listed above, the fourth in line ie Raj Roop Sagar, had two daughters and three sons, the elder daughter `Santosh` (Gone to Gulatis), the younger daughter `Veena` (Gone to Puris), The elder son `Prem Kumar Sagar` who is no more, and the younger one Lt. Col Vinod Sagar, who lives in Bangalore.

And the one in the middle is called :

SURENDRA KUMAR SAGAR

That's me

Surendra Kumar Sagar has two **sons** .. The elder one Kamal Sagar (owns the Total Environment Group of Companies) has three sons Aadi Nandan Sagar, Avi Vikram Sagar, and Abhi Veer Sagar… The younger son Nikhil Sagar is an American citizen who lives and works in Seattle,was married to an American MKMeador. Her father was the late Dr. Clifton.K.Meador, a famous Doctor and writer of several books including the bestselling `Fascinoma`. ... He lived in Nashville.

As you can see .. the girls unite the Sagar`s with the Gulatis, the Puris, the Handas, the Chadhas, the Khannas, the Kochhars, the Meadors... and so on.

And now, as promised, I come to the gentleman called Shri Ram Saggar.

Now I do not know How many sons and daughters, Shri Ram Saggar had, but one of his sons was named Kunj Lal Saggar.

And one of Kunj Lal Saggar`s sons was named Ram Prashad Saggar.

And one of Ram Prashad Saggar`s sons is a man called Balraj Sagar, and he is alive. He is an American citizen who lives in US and often stays in Nashville.

And finally I go back to the `Rock Island Line` ... and the two gentlemen in their eighties who watched the Gran Ole Opry ... are none other than Dr. Clifton K Meador (CKM) and Mr. Balraj Saggar (BS).

Indeed .. this is a small world .. not just in space but also in time.

And I am happy to say that the common denominator between the two gentlemen ... CKM and BS ... is that they were both reading my book : ``INTELLIGENT FIELD``

And I am happy to say that this gentleman called `Balraj Saggar` visited India recently, and he came to Bangalore, and he came to Flat number 221 at TIME 2, and had dinner with us.

I posted the story in part on Facebook.

A friend of mine called `Salim Sheikh` (we have not met) asked a question.

Why did Kishori Lal Sagar dropped a 'g'?.....In accordance with Newton's third law, the plane and the seat underneath the pilot provides an equal and opposite force pushing upwards with a force of 725 N (163 lbf). This mechanical force provides the 1.0 g-force upward proper acceleration on the pilot, even though this velocity in the upward direction does not change ...In straight and level flight, lift (L) equals weight (W). In a banked turn of 60°, lift equals double the weight (L=2W). The pilot experiences 2 g and a doubled weight. The steeper the bank, the greater the g-forces. Does it mean that Bhoj Raj Br was doing steeper bank transactions than the Kishori Lals?..Just kidding.. :)

I thought for sometime and then I wrote:

@ Salim Shaikh ... Good Question ... I`ll explain : Kishori Lal Sagar was an Engineer from I.I.T. Roorkee ( known as Thompson College of Engineering in his time). He was a fan of Newton, and hence retained the two g`s for quite some time. Later, in the nineteen twenties when Kishori Lal was in his thirties, he switched his allegiance to Einstein. He understood that in modern physics, we need to view the history of the universe physically as it happened, and as it is still happening, in a four-dimensional space-time, rather than in a three-dimensional space evolving with time. When Einstein became accustomed to the idea of space-time, he took it completely into his way of thinking. It became an essential part of his extension of special relativity to what is known as general relativity. In Einstein's general relativity, the space-time becomes curved, and it can incorporate the phenomena of gravity into the curvature. So, Kishori Lal could –

if he wanted to - actually have dispensed with both the g`s, but he decided to retain one `g` as a mark of respect for Newton.

Besides, he understood Quantum Physics, and knew that `Wave – Particle duality` was one of the central ideas of quantum Physics. As per the Copenhagen Interpretation of QP and utilizing Bohr`s idea of `Complimentarity`, he thought he was entitled to adopt the rule that he could use either a particle-like description of physical phenomena or a wavelike description. And he preferred the `Wave` to the `Particle`. As soon as he thought of `Waves`, he was reminded of the `Ocean`. Now, we all know that `OCEAN` = `SAGAR` ( with one `G`). So, his mind was made up, and one `g` had to go.

… … …

And here is another short story … a really short one :

Sometime in the 18th century there lived a man called ` `BHOJ RAJ SAGGAR` `. If it wasn`t for him, there would have been no `Total Environment` … and you would not be sitting/standing here, and listening to the story … `Interesting Small World`

…   …

EINSTEIN`S RELATIVITY AND DONALD CROWHURST`S SUICIDE

``There is no such thing as a `Causality Violation Paradox`, as no law of science ever gets violated. Its just that the law of science needs to be understood perfectly``

This short piece aims at reducing the misunderstanding between Scientists who support Einstein`s Relativity theories and some others who don`t.

Johan Prins is the author of the book ``THE PHYSICS DELUSION``. He is also a friend on Facebook. He is not in agreement with some aspects of Einstein`s Special Relativity, such as `Time Dilation` and `Length Contraction`, as well as the concept of `Space-Time` in Einstein`s General Relativity. Incidentally, there are a few more members of the club of scientists who oppose Einstein`s relativity. There was a man called Donald Crowhurst who was driven to madness after reading something that Einstein wrote. So much so that within a week of that `reading of that sentence` he committed suicide by jumping from his yacht into the sea.

This actually happened, details taken from a book `EINSTEIN`S MISTAKES` .. By Hans .C. Ohanian. :

On Tuesday June 24, 1969, Einstein Drove DC into madness. DC was a participant in the single handed around the globe sailboat race organised by the Sunday Times London. He had ample time to read in his yacht. He was reading Einstein`s book

``Relativity the Special and the General Theory`` and came across a para that blew his mind. Einstein proposed to test the simultaneity of two events – say two lightning strokes – at two widely separated points A and B by observing light signals from these events at the midpoint between A and B.

Einstein claimed: `That light requires the same time to traverse the path A to M as for the path B to M is in reality neither a *supposition nor a hypothesis* about the physical nature of light, but a *stipulation* that I can make of my own free will in order to arrive at a definition of simultaneity`

DC could not believe this. He knew that - because of the motion of the Earth – it was by no means self-evident that the speed of light relative to the Earth would be the same in all directions, and he thought it outrageous that Einstein would pretend to settle this question by *stipulation.*

In his logbook DC wrote : ``I said aloud with some irritation ` You can`t do THAT!` I thought `the swindler`. Then I looked at a photograph of the author (Einstein) in later years. The essence of the man rebuked me. I reread the passage and reread it, trying to get to the mind of the man who wrote it. The mathematician in me could distinguish nothing to mitigate the offending principles. But the poet in me could eventually read between the lines, and he (Einstein) read `Nevertheless I have just done it, let us examine the consequences`.``

But DC`s attempts to understand the consequences led him down a dizzying spiral into madness. For the next seven days, he wrote some 25,000 words at a furious rate, almost non stop. He wrote like a man possessed, frantically filling his logbook with a string of nonsensical pseudoscientific sentences, such as :

``I introduced the idea $(-1)^{0.5}$, because (it) leads to the dark tunnel of the space-time continuum, and once technology emerges from this tunnel the world will end (I believe about the year 2000, as often prophesied) in the sense that we will have access to the means of `extra physical` existence making the need for physical existence superfluous … And yet … and yet …if creative abstraction is to act as a vehicle for the new entity, and to leave its hitherto subtle state, it lies within the power of creative abstraction to produce the phenomena!!!!!!!!! We can bring it about by creative abstraction! He thought that he had discovered the means to impose his free will on the material world, and that he was thereby solving all the problems of mathematics, physics and engineering. Mathematicians and engineers used to the technique of system analysis will skim through my complete work in less than an hour. At the end of that time problems that have beset humanity for thousands of years will have been solved for them``

AND THEN THIS IS WHAT HAPPENED:

Toward the end, DC became obsessed with time, and he labelled each paragraph with the hour, minute, and second it was written. But in his writing frenzy he had forgotten to wind his chronometer, it had stopped. He restarted it on the basis of a rough assessment of local time according to the position of the Moon in the sky

Finally on July 1 at 11 hours 17 minutes, 0 seconds he wrote:

11 17 00 It is time for your move to begin … I have not need to prolong the game … It has been a good game that must be ended at the … I will play the game when I choose I will resign the game …

11 20 40 There is no reason for harmful …

He stopped in midsentence, at the bottom of the page. He ripped the chronometer from its bulkhead mount, and then - presumably at 11 hours, 20 minutes, and 40 seconds – he went on deck and threw the chronometer and himself into the sea. His trimaran was found six days later by a passing cargo ship, adrift and abandoned, with his logbook still lying on his desk, opened to the last page of his writing. His body was never found.

I will come back later on the Crowhurst story and the Author`s ( of the book `Einstein`s Mistakes`) view on the subject.

Meanwhile, here is another story. While being driven from residence to office here in Bangalore one morning, I happened to read an article entitled ``Fighting for Time`` by Graham Farmelo, in the journal `Nature` (May 2015 issue) borrowed from the British Library some days back. The article is precisely on the same topic which is discussed here , wherein Farmelo gives an account of Einstein`s clash with philosopher Henri Bergson, another member of the club of Scientists who considered Einstein`s Relativity theory as a Flawed philosophy. As soon as I reached office I sent a friend request to Graham Farmelo, and to my pleasant surprise, he accepted, and so he too is a friend on facebook.

BACK TO THE CROWHURST STORY:

Although Einstein`s book was the proximate cause of Crowhurst`s descent into madness, the stress of the preceding month was undoubtedly a contributing factor. (For more on this relating to the boat race… refer to the book `Einstein`s Mistakes). In his tormented state of mind he was ready to be pushed over the edge by any additional mental overload – and Einstein`s book gave him that push. Crowhurst was driven into madness by Einstein`s assertion that the equality of travel times by light signals in opposite directions is a matter of free will, but when Crowhurst explained that this was not permissible and that it was a swindle, he came close to the truth. The equality of these travel times is not a stipulation made by an act of free will, but a hypothesis which is experimentally testable and is found to be valid because the laws of nature make it valid. In treating this as a stipulation, Einstein made a conceptual mistake (as per the author of `Einstein`s Mistakes) – he adopted the right equality but for the wrong reason. In theoretical physics, the end does not justify the means. The theoretical physicist is not only expected to obtain the right results but is also expected to give us a coherent understanding on how to reach these results.

BACK TO THE CLUB MEMBER HENRY BERGSON :

In the words of Graham Farmelo: ``Henry Bergson chastised relativity theory for going beyond physics, to become a `flawed philosophy` that should be strongly resisted. He felt that human consciousness plays a crucial part in our knowledge of the Universe, so a complete account of time must reflect its subjective aspects. He debated the subject with Einstein, who strove to remove subjective elements from his theories. Einstein`s reply was terse to the point of rudeness. He said that there were only two ways of understanding time – psychological and physical – and the philosopher`s time did not exist. He believed that Bergson was confused and ignorant about relativity. Bergson was convinced that his opponent had not understood him. Bergson plainly did not comprehend basic aspects of relativity, so it is hardly surprising that this spat did nothing to make leading theoreticians reassess the theory. But it did some damage to

Einstein. In 1922 Einstein received the Nobel Prize, but the citation mentioned his work on the photoelectric effect, not relativity, and the Nobel Committee explained by referring to Bergson`s challenge to his relativity theory.

Another critic of Einstein`s Relativity theory was the physicist Herbert Dingle, who lamented that in general the scientist `understands what he is doing about as well as a centipede understands how he walks` ``.

Having conveyed the views of some critics of Einstein`s relativity, I ask the question: What can be that explanation which can bring about an agreement between the supporters of Einstein`s relativity, and those that do not support it?

Let me attempt to answer that question.

Before I proceed further, I reproduce here a small extract on the subject .. `Of The Concept Of An Object Of Pure Practical Reason` .. from Immanuel Kant`s book `The Critique of Pure reason` :

`` Now, since the notions of good and evil, as consequences of the a priori

determination of the will, imply also a pure practical principle, and therefore a causality of pure reason; hence they do not originally refer to objects (so as to be, for instance, special modes of the synthetic unity of manifold of given intuitions of one consciousness) like the pure concepts of the understanding or categories of reason in its theoretic employment ; on the contrary they presuppose that objects are given; but they are all modes (modi) of a single category, namely that of causality, the determining principle of which consists in the rational conception of a law which as a law of freedom, reason gives to itself, thereby a priori proving itself practical. However, as the actions on the one side come under a law, which is not a physical law, but a law of freedom, and consequently belong to the conduct of beings in the world of intelligence, yet on the other side as events in the world of sense they belong to phenomena; hence the determinations of a practical reason are only possible in reference to the latter, and, therefore in accordance with the categories of the understanding ; not indeed with a view to any theoretic employment of it, i.e, so as to bring the manifold of (sensible) intuition under one consciousness a priori ... But only to subject the manifold of desires to the unity of consciousness of a practical reason, giving it commands in the moral law, ie , to a pure will a priori`` `

After five or six readings of the above para, this is what I thought it implied:

` `Before we make our judgment of whether something is `correct ` or `incorrect `, we must ask ourselves under which law we are making the judgement, whether it is the law of freedom (of will), or it is a physical law (law of Physics). The former from considerations of `Pure Practical Reason` corresponds to the conduct of beings in the `world of intelligence`, and the latter from considerations of Pure Reason (Theoretical) would mean that any act performed is in the world of sense, ie just a phenomena ie. merely the motion of atoms and molecules in accordance with the laws of causation` `.

Having said that. I start with Einstein`s wonderful claim referred earlier:

Einstein claimed: `That light requires the same time to traverse the path A to M as for the path B to M is in reality neither a *supposition nor a hypothesis* about the physical nature of light, but a *stipulation* that I can make of my own free will in order to arrive at a definition of simultaneity`

Each one of us, after developing a thorough understanding - commensurate with the capacities of our brains – of the subject based on logic and rational thinking, can make stipulations of his own free will in order to arrive at a definition of simultaneity.

Somebody can than study all the `stipulations` (Inputs) available to him (or her), and pick all the common denominators, and add his own `stipulation ( if any)` in such a way so as to arrive at a final `stipulation` which is as far as possible having the best chance of being acceptable to one and all.

Keeping in mind what has been stated above, If I am that somebody, if I combine the two laws referred above .. viz the law of freedom (of will), and the physical laws (laws of physics), and if I use both hemispheres of my brain - the left dedicating itself to logical and analytical thinking, and the right to an intuitive and holistic understanding – with its limitations, this is what I have to say:

` `An instant of time is indeed a well-defined instant of time for all observers who are aware of that specific instant of time.

The duration of time - as actually happens in a universal frame of reference - between two specific and well defined instants of time `T1` and `T2`, is also a fixed and well defined duration of time `t` for all observers who are aware of each of these two specific instants of time, provided either that they are stationary in one place, or if they are moving in a space ship, they do take into consideration the effect of `Time dilation` (caused by moving at high speeds comparable to the speed of light) in their calculations. Thus the travelling observer `O2` if he has the means of measuring time `t1` as experienced by him during the time between the two specific instants of time T1 and T2, and is also aware of the speed at which the spaceship is moving, then `O2` can calculate the actual duration of time `t` as actually happens in a universal frame of reference, or as measured by a stationary observer `O1`, by making use of the `Lorentz transformation`. With all this, there is no such thing as a `Casualty Violation Paradox`, as no law of nature is violated, its just that the law is understood better.

Now there are all sorts of complexities involved when we take into consideration the motion of planets, stars and galaxies, and that each of these turns out to be a spaceship of some kind. Thus a `Star System` turns out to be an observer and the `Galaxy` is his spaceship. In the same way a `Planet` is an observer and the `Star System` is its Spaceship. In this scenario, it is impossible to visualize a stationary observer on the planet Earth, as the planet `Earth` then turns out to be his spaceship. Hence for all practical purposes, there is no such thing as a stationary observer in the universe, and hence for all practical purposes there is no such thing as an absolute time that goes `tick tock` all the time as there exists no preferred observer to measure it.

When two observers meet at some point and shake hands, then, regarding the event corresponding to the shaking of the hands at the point corresponding to the junction between the two hands (if it happens at a certain time "T"), we can very well say that time 'T' is a fixed instant of time (albeit in someone's imagination). But the interval of time between the big bang and time T is not the same for the two observers or for their constituent parts, the trillions of atoms and molecules. This interval of time is different for each and every one of these constituent parts and depends on their travel schedules and the speeds at which they were traveling in their journeys through the cosmos via supernovas and so forth. It is also dependent, of course, on the number of times Lorentz transformations were applied to them.

However there is no stopping us in our mind if we wish to imagine ourselves as a mind in a universal frame of reference, perhaps `stationary` and stranded somewhere in interstellar space (not moving with any star or galaxy) and doing nothing except measuring time. In such a state of `Mind` we can very well consider time to be `Absolute Time going `Tick Tock` all the time.

Now consider the speeds involved in the motion of planets and stars and they are nowhere near the speed of light:

Earth travels at about 108000 km per hour, taking about 365.25 days to complete one lap of about 970 million km around the sun. Our Solar system travels at about 792,000 km per hour and takes about 225 million years to complete one lap around the centre of our milky way.

Spaceships as designed by human beings are unlikely to attain these speeds, however even if we assume an observer travelling in a spaceship moving at a million kms per hour, after travelling for an year between two specific instants of time, the quantum of time dilation is likely to be less than a second, so it can hardly be called a paradox of any kind.

Hence the conclusion:

`There is no such thing as a `Causality Violation Paradox`, as no law of science is violated, its just that the law of science needs to be understood perfectly. The cause itself needs to be understood properly before deciding if the corresponding effect is a paradox or not.`

If Donald Crowhurst had read this piece, perhaps he may not have committed suicide. Of course, it is also true that, had Donald Crowhurst not committed suicide, I may not have written this piece.

… … …

COLONIZING THE GALAXY

The number of stars in the Milky Way is so large that the probability of at least 100 Technological civilizations at least as intelligent as the one on the planet Earth is reasonably high as worked out below:

(Refer `Cosmology` ….. By Edward R. Harrison – The chapter on `Life in the Universe`)

If N = Number of stars in the galaxy (Milky Way), Then the number of Technological Civilizations (TCs) that may have existed/still existing is given by:

$$n = N*P1*P2*P3*P4*P5*P6*P7*$$

Where P1 is the fraction of stars in the galaxy similar to the Sun ie not too blue, not too red and not members of closed binary systems. Reasonable estimate 1/10 ie 0.1.

P2 is the fraction of sun like stars having earthlike planets. Optimistic estimate 1.0 and conservative estimate 0.1

P3 is the fraction of such planets occupying a habitable zone, ie neither too close (like Venus), nor too far (Like Mars) from the parent star. Reasonable estimate 0.1

P4 is the probability that life originates in a unicellular form. Reasonable estimate 0.1

P5 is the probability that it evolves into complex multicellular organisms like mammals. Many optimists may consider P5 to be close to unity but recognizing the hazards and traps involved a conservative estimate is 0.1.

P6 is the probability that life develops intelligence. Again many optimists consider this inevitable but ERH again takes a conservative approach. He feels that the environment must in some way administer rude shocks in the right way at the right times to the right species, so that natural selection will favor the development of large brains and the probability of this happening is very small - - about 0.1

P7 is the probability that intelligent life discovers Science and develops an advanced technological civilization. Here again an optimist would take P7 as almost unity but a pessimist would take a different view. Science was not discovered by the cultures of Africa, America, China, India, and Japan etc.Science arose because of accidental and improbable circumstances that existed in Greece and later in Europe. Efforts of a few individuals, who rejected the magic of gods, can be considered as the first steps that created Science. Later it was revived in Europe in the face of organized hostility. Again a conservative estimate of 0.1 is assumed for P7.

Based on above estimated values of the factors P1 to P7, the likely number of TCs (in the Milky Way only and not the Universe as a whole) that may have existed sometime in the past, inclusive of those which are still existing varies from about 10 million(Optimistic estimate) to about 10000(Pessimistic estimate).According to ERH this number is closer to the lower estimate.

How many of these TCs still exist today?

If this number is denoted by n1,its value is equal to P8*n. What is P8?

There are many many ways in which a TC can collapse, such as cosmological collisions (Asteroid hits in case of Earth), Climatic and other conditions no more suitable for continuance of life, but none of these has a greater probability than Self destruction by aggressive approach.Developement of Science creates new hazards and the chances are that most TCs are short lived. Conflicts, Wars, Terrorism, Misadventure, and above all the possibility of Nuclear Wars, gives a low value of 0.1 (optimistic) or 0.01(Pessimistic) to the factor P8.

And so out of 10000 TCs that have arisen in the galaxy, perhaps between 100 and 1000 are still surviving.

How many of these TCs which are still surviving, likely to survive permanently?

If this number is denoted by n2, its value is equal to P9*n1. What is P9?

Two objectives must be fulfilled in order to survive permanently (at least as long as the Stelliferrous age):

GALACTIC COLONIZATION: The TC must survive about 50 million years (by avoiding self-destruction) to achieve this objective, to develop interstellar space travel, to develop fusion power and other technologies to construct large space vehicles that can travel at one-thousandth the speed of light, as well as have its own biosphere and containing a social unit of tens of thousands of people to enable 10000 years of travel time at a stretch, by halting at one destination for 10000 years before embarking on the next journey. In this way it is possible to diffuse outward @ 10 light years per 20000 years and colonize a substantial portion of the galaxy in 50 million years (which is not even 0.5 percent of the present age of the Universe).

GALACTIC SELECTION: The TC must understand and implement properly a certain law called ` THE

BIOGALACTIC LAW'. Given below is a very interesting para from ERH`s book referred above, on this subject:

`` THE BIOGALACTIC LAW: We are the outcome of natural selection and to the operation of this law can be attributed the fitness of the human body and brain. Presumably this is true also of the life on other planets that attains a state of advanced intelligence. When a civilization gains control of its planetary environment, the evolutionary game changes, however, and new rules determine what is fit and unfit. Natural selection now operates on a planet-wide scale according to a biogalactic law that will be referred to as galactic selection. This speculative law of Galactic Selection states simply :`Intelligent life forms that are destructively aggressive do not colonize the galaxy'. This law operates conceivably in two modes; the first is unconscious and automatic, and the second is conscious and deliberate.``

The two words `do not` in the above definition of the Biogalactic law probably correspond to the five words `will not be permitted to`.

It goes without saying that the above objectives are indeed very difficult to achieve and even a very optimistic value that can be accorded to the factor P9 cannot exceed 0.1.

Hence the conclusion:

The Technological civilization of the planet Earth may have reached the top hundred in the Milky Way, but it may not reach the top ten if it self-destructs before that.

....

ARE THEY `BLACK` ...
ARE THEY `HOLES`

Whatever they are .. they are already here even when the universe is still in its `infancy` .. A time will come -from 10^{40} years since the big bang to 10^{100} years since the big bang - when they will inherit the universe completely

This subject is dealt in two parts. Part One corresponds to the part that I understand somewhat, which is in fact a portion of what is understood so far, and Part Two corresponds to the Part which I do not understand, and a portion of this part two is actually not understood by anyone till now.

WHAT I DO UNDERSTAND ... SOMEWHAT ... ABOUT BLACK HOLES

MY UNDERSTANDING OF BLACK HOLES:

To begin with.. Are they black ? :

Yes indeed they are black. Of this there is no doubt whatsoever. Light could not escape from them, so we cannot see them. We may have an approximate idea where they might be located, from the way they attract material towards them, but when we look in that direction – assuming there is nothing in between – we find pitch darkness, as black as ever. Another reason to pin point on their blackness is their temperature, which is almost close to the minimum possible. As you know a cold metal bar in a dark room looks black and the lower is it`s temperature, the blacker it looks.

Our parent black hole – of the Milky way – is real cold with temperature of about 1.5 x 10 ^ -14 degrees K.

Are they holes ?

They cannot be ... A black hole is `something` truly massive ie something `solid` inside interstellar space which is literally `nothing`. if we pierce a hole in a solid or something...then thats a hole... and there is nothing in that hole...which means a `hole` is `nothing` in `something`... then how can `something` in `nothing` - which is the reverse of `nothing` in something`- be also considered as a hole ?

On the other hand...if a black hole is indeed a gigantic hole ... then how is it massive? ... and how is it pulling things? ... and how is it preventing light to escape from its clutches ? ... and how is it creating `curved space` arround it ?

Or should we just call it a `Singularity` ... and tell the world that the laws of science do not work here ... and be done with it ?

Where do they come from ... these black holes ?

From Supernovas of course... Recall my autobiography, and my experience in a dying star ... When the collapse of the star took place, there were three possible scenarios I could be a part of :

Scenario one : Complete expulsion of all material into space ... a victory for entropy.

Scenario two : Expulsion of most of the material into space, and retaining some of it in the heavily contracted central mass and becoming a Neutron star ... a partial victory for entropy and a partial victory for gravity.

Scenario three : Expulsion of some material into space, and retaining most of it in the heavily contracted mass via a massive implosion and becoming a Black Hole ... a victory for gravity.

As I have lived on to write my autobiography... the probability is very high that I was part of scenario one... some probability that I was part of scenario two... small probability that I escaped into space in scenario three ... and zero probability that I went into a black hole in scenario three. Many of my friends however managed to get there and are still entrapped in the black hole. I have so far not been able to establish contact with them... I am waiting for Hawking radiation ... and I`ll take my chances thereafter.

How big are they as compared to how big they were before they became black holes:

Just about as small or as big in size so that the density of material becomes so staggeringly high and the gravitational pull so staggeringly huge, that to escape from it you need to exceed a speed of 300,000 kms per second, ie exceed the speed of light... which means even light cannot escape their clutches. This size is also called the `Schwarzschild radius` of the body. The Schwarzschild radius for our Sun is about 3 kms, which means if our sun were to become a black hole it would just be a sphere of 3kms radius. And if our planet Earth were to become a black hole, it will be a sphere of just a couple of cms in diameter. All the mass of the black hole gets squeezed inside this critical radius. Curved space completely surrounds the black hole, and pinches it off from the rest of the universe. Matter can fall into this, but not escape from it, not even light. The Schwarzschild radius of an object can be worked out as equal to $2G*M/c^2$, where G is the gravitational constant, M is the mass of the object and c as the speed of light.

What properties define a black hole? :

Only three, its mass, its charge, and its angular momentum (rotation). So, a black hole can only be in one of the four categories...viz uncharged and non-rotating, charged and non-rotating, uncharged and rotating, and charged and rotating.

What Kind of laws of science govern inside or arround black holes when light is unable to escape from them ? :

Nobel Laureate Subramanian Chandrasekhar considers Black holes as perfect macroscopic objects. He wrote:

`` *The black holes of nature are the most perfect macroscopic objects there are in the universe: the only elements in their construction are our concepts of space and time. And since the general theory of relativity provides only a single unique family of solutions for their descriptions, they are the simplest objects as well* ``

But there is nothing simple about what happens inside or around Black holes within the so called `Event horizon`.. the surface area around the black hole from which nothing can escape .. not even light, not even information.

It was Einstein`s theory of General relativity which resulted in the concept of a black hole as something possessing an enormous gravitational pull so powerful that whatever entered within the event horizon was forever trapped and completely captured .. not to mention `crushed` with no chance of escaping.

Most black holes are products of Supernovas. In the final moments of a dying massive star (several times as massive as our sun), there is a fight between Entropy and Gravity. If Entropy wins, all the matter in the exploding star gets out and nothing remains inside.

If it's a close fight such as a short head victory for Entropy a portion (somewhat major) escapes out and a portion remains which is called a neutron star. And if Gravity wins, the star becomes a black hole.

When Quantum Mechanics was understood (sometime in the mid 20th century), people started thinking that the `information` - ie. The material sucked into the black hole - had to be conserved, and that it should be possible to retrieve the same at a later date.. but perhaps in a different form altogether.

Then in 1970 the Scientist Stephen Hawking proposed that black holes were losing mass and would eventually evaporate and in doing so take all traces of what fell into them. Thus the information would not be conserved. It would be vanished.

What ultimately happens to the information is unclear and this aspect is known as `The information paradox`.

How to resolve the Paradoxes in connection with Black Holes:

There are a whole lot of other paradoxical situations that arise consequent to the fact that light cannot escape from the black hole, such as what does a person trapped inside a black hole (assuming he is somehow alive though in pieces) see when he looks outside : Does he see the future unfolding itself all of a sudden ? (I wish he calls me up and tells me who is going to win the next Kentucky Derby).

How can that be ?... What does he actually see ? …. This is so difficult to believe that sometimes I think Black Holes can`t exist.

Indeed they do exist … they have been observed for decades .. observed by deduction of course as the invisible members of curious double stars, where the visible member's movement around the invisible one provided enough information to prove that the invisible one was much more massive than a white dwarf or a neutron star and was thus a black hole. There is one in the centre of our own galaxy .. the Milky way… a really massive one, perhaps millions of time as massive as our sun. This is borne out otimese fact that all the stars with all their family members (the planets), are orbiting them. In fact all the galaxies must have a massive black hole at their centres.

Then there are the exotic and immensely unbelievable paradoxes involving the interconnectedness of universes via things known as `wormholes`.

This is too much.

But even some top-level scientists believe that a certain `Space-time` (derived from some interesting mathematical features) can be thought of as representing a time-evolution of two different universes that become connected by a `wormhole`, which subsequently `pinches off` in a singularity.

As I said earlier … This is too much for me!

Now …how do we explain and take care of these paradoxes?

Simple … We apply the same principle as we do while discussing the `Twin Paradox` … viz that there is no such thing as a paradox, if the physics concerning the cause is perfectly well understood, and the effect happens in accordance with expectations. We call something as a paradox only if we find that a physical law is violated. But in case of the `Twin Paradox` the physical law has NOT been violated. Its just that we have a better understanding of the physical law, the law to which a Lorentz transformation has been applied, to make it in conformity with what actually happens in the world. In short, the transformed Physical Law has not been violated at all

The trouble is, we do not yet understand the physics relating to what happens in and around black holes perfectly well. There appears to be a problem in understanding what happens in extreme situations encountered in black holes – a sort of a mismatch in the outcome when we apply Einsteins General Relativity ..with that when we apply Quantum Mechanics. Both these theories appear to break down at these extreme points and it becomes necessary to look towards a new theory of `Quantum Gravity`. In short .. in the physics of `Quantum Graviy` we need to look for an appropriate change (a sort of transformation) to these two theories such that the outcome is similar in both.

Till date there is no such consensus on a perfect understanding of what this quantum gravity is. But whenever it happens I guess each and every paradox

…. …. ….

OSCAR OF THE MIND

Is this an Oscar which I see before me ?

With its base towards my hand

Come, let me clutch thee

What ! I have thee not and yet I see thee still

Art thou not fatal vision sensible to feeling as to sight ?

Or art thou but an Oscar of the mind ?

A false creation, proceeding from the heat oppressed brain

Or is this a universe different from the one

where the coveted prize yours truly won

For acting his own part in the Spielberg movie ` `Seminar held in a parallel universe` `

based on his own book `Six Words`

And the Oscar delivered to me by none other than JL + PC

And I gave my acceptance speech thus :

` ` ` This indeed is like a Cosmic Religious Feeling. For this I have to thank many

Beginning with you Albert Einstein for your inspiration, and for your perfect understanding of God, never mind if he is not anthropomorphic.

Thank you dear Erwin Schrodinger for that Arithmetic Paradox , `The oneness of the mind`, never mind if the consciousness is in the singular.

Thank you, John Wheeler for that Self excited circuit, that observer created reality, that delayed choice experiment that allowed me to see the past.

Thank you Rene for the quote `I think therefore I am`

Thank you Kant for the consciousness unity. Never mind the complexity `A priori`

Thank you Rein court for `Squaring the circle - beyond the mind`

Thank you Paul Davies for your conviction `God is a mind, of the infinite kind`

I think it is a `Travelling Cosmic Mind`

Thank you Eugene Wigner for your wonderful idea that has enabled my consciousness

To collapse the wave function of this Oscar so I can hold it in my hand

And tell the World, we are all one and of the same kind

Yes indeed, `We all have the same mind `

The Six Words ```

....

SHAKESPEARE AND THE 2020 PRESIDENTIAL ELECTION

Joe Biden had a comfortable lead over Trump in most of the polls. But there was a distinct likelihood that the November election would be plagued by voter suppression, foreign interference, disinformation and a bitterly contested supreme court vacancy. Without doubt this was a recipe for chaos specially as all this came after a disastrous year in which the US was considerably shaken by the deadly coronavirus pandemic, economic collapse, and widespread racism. It was even being dubbed <u>"the election that could break America"</u>.

The expectation was that Biden would win in a tight race and Trump could refuse to leave office, calling the election as `Rigged` And it all turned out to be just as expected. Biden won the November 2020 Election in US in a tight race as was expected. Trump kept refusing to step down, also as expected.

Perhaps Biden had visions and certain thoughts came to his mind, thoughts like:

``To be thus is nothing, but to be safely thus

Our fears in Trump stick deep,

And in his capacity to create chaos reigns that which would be fear'd.

Tis much he dares, and through that tribal support of his base,

He hath a confidence, that doth guide his Valor.

To act in safety. there is none but he, whose being I do fear.

And with all this, will my chances get reduced?

He incited the voters,

When first they put the Name of Potus upon me,

And bade them speak to him.

But then with Proper sense, they did ignore him.

And upon my Head they plac'd a flawless Crown,

And put a barren sense in his griping

So, no chance of him succeeding,

and coming in my way``

But Trump would still be in control till January 2021, with a military aid carrying the brief case containing the nuclear football always by his side.

An article by BOAS (Bulletin of the Atomic Scientists) of extreme significance and very disturbing to the thinking mind appeared in January 2021. Reproduced below is a small extract from the article:

``President Trump is followed 24/7 by a military aide that carries the "football," the briefcase that holds all he would need to order the immediate launch of up to 1,000 nuclear weapons, more than enough megatonnage to blow the world back into the stone age. He does not need the approval of Congress or the secretary of defense. Shockingly, there are no checks and balances on this ultimate executive power.

President Trump took the nuclear football with him to Walter Reed Medical Center, where he received treatment for COVID-19. According to Trump's doctor, the president's blood oxygen levels had dipped. And this, according to <u>independent health experts</u>, can impair decision-making ability ... and so on `` ``

So, The President has the sole authority to initiate a nuclear attack and there are no safeguards to verify his sanity.

And in the case of The Donald, his sanity - mental state - was questionable even when he was not having a Coronavirus attack.

Perhaps the Donald had visions and may have had thoughts like:

``*Is this a briefcase with a Nuclear Football inside it, which I see before me,*

The handle toward my hand? Come, let me clutch thee.

I have thee not, and yet I see thee still.

Art thou not, fatal vision, sensible

To feeling as to sight. Or art thou but

A briefcase of the mind, a false creation,

Proceeding from the heat-oppressed brain?

I see thee yet, in form as palpable

As this which now I draw.

Thou marshall'st me the way that I was

going and such a weapon I was to use.

Mine eyes are made the fools o' the other senses,

Or else worth all the rest. I see thee still,

And on thy top and sides fire and radiation,

Which was not so before. There's no such thing:

It is the bloody business which informs

Thus to mine eyes. Now o'er the full world

human civilization will be dead, and wicked dreams abuse

The curtain'd sleep; TRUMP-CRAFT will celebrate

humanity`s end. A bell rings.

I go, and it is done: the bell invites me

Hear it not, for it is a knell

That summons thee and Xi to heaven, or to hell``.

....

The second impeachment of Donald Trump, the 45th president of the United States, occurred on January 13, 2021, one week before his term expired. It was the fourth impeachment of a United States president, and the second for Trump after his first impeachment in December 2019. He was impeached by the House, but in the Senate trial, a two-thirds majority was required to vote for his removal as President.

Now, there were 100 Senators, 53 Republicans and 47 Democrats. There was only one truth. Did the senators arrive at a consensus on what it was?

That is the question.

I am thinking. Assuming all the brains were wired to a computer and a suitable contraption was used, making it impossible for the senator to remember if he or she was a Democrat or a Republican and assuming further that all 100 senators were perfectly well-informed, intelligent and honest, then the probability that the Democrats arrived at one truth and the Republicans arrived at a different truth was $= (0.5) \wedge 100$, just about as low as that of an aircraft being assembled by a tornado striking a junkyard.

All but one Republican voted against Trump's removal, and Donald Trump was acquitted.

I call it nothing other than 'Tribal Loyalty.'

Now, it cannot be that the Republicans were not well informed on the facts of the case. It cannot be that they did not understand the law of the land. It cannot be that they were not intelligent enough to be able to analyze and understand what the truth was.

But were they honest? That is the question. No, They were not honest. So, I call it nothing other than 'Debauchery'.

… … …

IS THIS OUR FINAL MILLENNIUM

Some days back while driving from `TIME` to `IMAGINE` ie from Residence to office, I was listening to Stan Getz on the car CD player. Before playing a number called COBA, he said a few introductory words.

And I add a few more.

Now, COBA is a small town in Mexico. Butterflies from North Mexico and South Mexico and several other parts of the world frequently assemble in this place. In fact they have their annual convention here. Looks like they carry out regular research on how to avoid getting extinct and take some important decisions to that effect during these conventions. Whether there is climate change or nuclear wars, they will ensure they will NEVER get extinct.

Nor will `Termites`.

Some years back .. again while travelling from `Time` to `Imagine` I was reading a very interesting short story called ``RETREAT FROM EARTH`` by Arthur C Clarke about Termites and their interactions with Aliens from outer space.

And I add something to that.

Now these Highly Intelligent Aliens from a certain planet orbiting a certain star .. I don`t remember the name .. have been trying to figure out how to venture to come to the planet Earth and try to colonize it. They are not at all afraid of us Human Beings, they can easily conquer and enslave us or maybe eliminate us .. whatever. But they are terribly afraid of these `Termites`. The

Aliens have tried hard to communicate with the `Termites` for obtaining permission to colonize the Planet .. but each time the Termites tweeted back ..``DON`T MEDDLE WITH HUMANS``.. OK?

In fact I can tell you .. after the human race gets extinct, there will only be two possibilities .. either all the souls will get stranded in interstellar space .. or they will go into Termites.

For Further listening .. about Butterflies` ... listen to Stan Getz` version of `COBA`

For further reading .. about Termites and their interactions with Aliens .. Refer Arthur. C Clark`s short story `` Retreat from Earth``

For further reading on How to prevent our extinction and going into Termites, read my book : `IS THIS OUR FINAL MILLENNIUM`

... ...

IMAGINARY CONVERSATION

Eugene Wigner: Recall your words, John, .. "Of all the strange features of the universe, none are stranger than these: time is transcended, laws are mutable and observer-participancy matters". What is the significance of these words?

John Wheeler: I guess my philosophy of science is much more radically relativistic than Einstein's. I would make all physical laws relative to observers, which means creating them by our mere existence. Life may even create physical laws by conscious decision. The universe I believe is a self-excited circuit. Starting small, the universe grows (loop of U) and in time, gives rise to observer-participancy, which in turn imparts 'tangible reality' to even the earliest days of the universe. Recall the *'Delayed-choice experiment'*.

EW: Yes, I remember your thought experiment. But isn't that a violation of our notion of causality?

JW: It is in this sense that quantum mechanics does strange things to what we call causality if we examine it with sufficient care. Hence, it is in this sense that there is no such thing as a causality violation paradox. It is just that we need a better understanding of the laws of nature. At a certain level, there must be a transformation in the law. We do not understand the mechanism of the transformation, but we do know that a transformed law of science is NOT violated.

But why do you ask these questions?

EW: I was wondering how far can we go back in the past? For example, if this Omnipresent cosmic mind—Omnipresent in space as well as time—decides to go back to the past, say to the year 1900, will the events of the 20th century be the same as they actually happened in this universe?

JW: No way. The probability of that would be just about as low as the probability of an aircraft being assembled by a tornado striking a junkyard.

EW: So no WW1 and no WW2?

JW: Maybe not. Or maybe there'd be four world wars. What do I know?

EW: On what would that depend?

JW: On the visions and ideas transmitted to the human minds of the past, based on the experience gained and the inputs received via the information in the intelligent field of the intervening period between the year 1900 and the present, by the traveling cosmic mind before it decides to go back to the year 1900.

EW: So if it turns out that there is a nuclear war in the very near future, followed by a massive chain reaction, followed further by a nuclear winter and the Doomsday clock finds itself at just a few seconds away from midnight and human civilization on the verge of extinction, and if the traveling Cosmic mind decides to travel back to the present moment, what would the visions and ideas required to be transmitted to the intelligent human minds of the present be?

JW: Good question. I believe what you are trying to say is, that the damage is not yet done, The Russia – Ukraine war has not yet turned nuclear, what should be done to prevent such a catastrophe?

EW: Yes, precisely. There is still time .. but the time is running out? What should be done?

JW: The Four world leaders .. of Russia, Ukraine, USA, and China must come together and adopt a unified approach to address the conflicts of the world.

EW: Where should they meet?

JW: Perhaps Elon Musk should invite all four of them for a free ride in his spaceship, the atmosphere would be quite congenial for an intellectual discussion and may possibly lead to a positive outcome.

EW: In short, SWITCH OFF the deep states of the world and SWITCH ON the good times for the lives on the planet.

JW: Yes, you are right.

EW: What if they keep fighting and don't agree on anything.

JW: Divert the spaceship to the nearest blackhole.

… … … …

QUIZ TIME

SIXTEEN QUESTIONS - Q1 TO Q16 - ASKED IN THIS SHORT PIECE

(For answers go to the piece entitled: Answers to Questions)

<u>AN EXERCISE IN SCRABBLE (For Q1 and Q2)</u>

Some years back I posted on Facebook regarding the maximum number of points that can be scored in Scrabble in a single turn. I thought I was bragging when I mentioned that it was possible to score 427 points by making the word `Quizzical` based on a particular setting specially arrived at to permit placing that word and get 427

There was some discussion on the subject with a friend `Roy Tomes` who suggested that scoring `1304` was a possibility with a contrived setting. I asked Roy Tomes not to reveal that word or the setting till I give up, to which he replied that he is not aware of that word, he only knows that 1304 is probably the World Record.

After some exercise I found a word that can fetch a score in excess of 1100 but could not find the word that gets 1304.

The assumptions I made were that:

1) It must be a 15 letter word that is placed on the topmost row which has 3 `triple word` score spaces in the 1st, 8th, and 15th spaces , and two `double letter` word scores in the 4th and 12th spaces.

2) 8 of the 15 letters must already be in place, and none of them should occupy the 1st, 8th, and 15th spaces, to enable the `triple word` space to be occupied by the new letters to get maximum advantage both in the `row` word as well as in the `column` words. Which means the 15 letter `row` word score will get multiplied by 27 times, and the three `column` words scores would each be multiplied by 3 times.

3) The fifteen-letter word must accommodate the criteria that the eight letters already placed must form two or three words, and the contrived setting should accommodate this aspect.

4) Either or `both` the 4th and `12th` space should have .. if possible ...one or more of the big point words such as `z`, `q`, `J` or `x`.

After a detailed study I arrived at a certain 15 letter word that fulfils the above conditions.

I will ask the reader to guess the word and then calculate how many points he can score by using that word.

Here`s the setting on the board that was necessary to arrive at:

The existing 8 letters in the topmost row are forming three words ..`MATER` placed in the 3rd to 7th spaces, 'I` in the 11th space and `IN` placed in the 13th and 14th spaces.

The existing word in the first column just below the 1st space is `EVOLUTION`, so you need to decide on which letter to be placed on the 1st space which when added to `EVOLUTION` at the top forms an acceptable word.

The existing word in the 8th column just below the 8th space is `SLANDER`, so you need to decide on which letter to be placed on the 8TH space which when added to SLANDER` at the top forms an acceptable word.

The existing word in the 10th column just below the 10th space is `ANGUISH``, so you need to decide on which letter to be placed on the 10TH space which when added to `ANGUISH` at the top forms an acceptable word.

The existing word in the 12th column just below the 12th space is `ONE`, so you need to decide on which letter to be placed on the 12th space which when added to `ONE` at the top forms an acceptable word. There may be many such letters possible which form an acceptable word in this case, but we need to choose the one letter which fits in with the main 15 letter word.

The existing word in the 15th column just below the 10th space is `RUMBLING`, so you need to decide on which letter to be placed on the 15TH space which when added to `RUMBLING` at the top forms an acceptable word.

But how did we arrive at the setting?

This is explained below:

Consider two players `Pat Boone` and `Bernardine` playing Scrabble;

Pat Boone plays first and uses 5 letters to make `WATER` placed left to right, with `W` on the `double letter space` and `R` on the `double word space` right in the centre of the board. He scores 24.

Bernardine plays next and uses all seven letters to make `SLANDER` (placed up – down) where the blank is used as `N` and gets 66, including 50 for using all seven letters.

Pat Boone uses three letters `B` `E` and `D` and makes two words BED (left right) and BE (up down) where B and D are placed on a double letter score spaces. He gets 18.

Bernardine again uses all seven letters to make `ANGUISH` (up down) and another word `BEDS` (left right), with `A` and `I` coming on the triple letter score space. She gets 15+ 7 + 50 = 72 points.

Pat Boone uses 4 letters to make `HARP` (up down) through `A` of the word WATER.

Bernardine makes `HARKEN` (left right) starting with `H` of `ANGUISH`

Pat Boone uses all his seven letters and makes `RUMBLING` (up down) through `N` of `HARKEN`. He scores 48 + 50 = 98.

And then Bernardine makes `OFFER` (left right) ending with `R` of `HARP` and `WE`

Pat Boone makes `ION` (up down) through `O` of `OFFER`...to score 9

Bernardine uses six letters to make a nine letter word `EVOLUTION` (up down) over `ION` ... to score 13

Pat Boone makes `OPAQUE` (left right) starting with `O` of evolution and gets 34

Then its `MOSAIC` (up down) by Bernardine though `A` of `OPAQUE`

`MATER` (left right) by Pat Boone starting with `M` of `MOSAIC`

`ANGLER` (left right) by Bernadine through `A` of `SLANDER` and `G` of `ANGUISH` using blank as `L`.

A few small words `OR`, `NO`, `IDOL`, `NON`, `ADO` and `ONE` and thats it.

We have reached the contrived setting

Readers can place the Scrabble board in front of them and place the letters in accordance with the sequence given above to arrive at the setting.

It is a valid setting... there is no question about that

It is now Bernardine`s turn to play

And she has the following seven letters with her: `E`, `I`, `Z`, `D`, `G`, `A` and `L`

What exactly is the 15 letter word that Bernardine will make, along with several other words to get a maximum score

Guess the word ... Q1

What is the total score achieved? ... Q2

In the process I discovered a new game on `SCRABBLE` which a single player can play alone.... Or it can be a number of players in a team

And he (or the team) has the entire set of tiles (102 of them) available to choose any seven for each turn of the play.

What is the objective?

Well, the objective is to play the game and arrive at a valid setting, then select your seven tiles, and play a single turn and make valid word/words to score the maximum number of points in just one turn.

And that is your (or the team`s) `SCORE` in the game.

The next player (or the next team) then plays the game and tries to achieve a better score.

Any score less than a thousand points, is not good enough.

<u>THE LAST SESSION OF THE SEMINAR (for Q3)</u>

After lunch on the last day, the delegates started returning to their seats.

The twenty-five who were in the Dais took their appointed seats in the front row of the auditorium. Classical music played in the background. The giant screen in front displayed images of spellbinding beauty. Each image was displayed for about a minute and then gradually disappeared in about fifteen seconds and the next one appeared. Against each image was a caption.

The number of words in each caption corresponded to an alphabet....

One word corresponds to `A` two words correspond to `B`..and so on.

The number of dots at the end of each set of captions corresponds to number of captions in each set of captions. This also equals the number of letters in that word.

Guess the Six Words.

.. ..

<u>THE FIRST WORD:</u>

My body functions as a pure mechanism according to the laws of nature, yet I am the person who controls all the motions.

It's all in the numbers.

..

THE SECOND WORD:

Note

The personal self equals the omnipresent eternal self`` is the grandest thought

The concept of the omnipresent mind leads us towards a cosmic consciousness.

...

THE THIRD WORD:

All religions are created by laws of science

And

In the eyes of the quantum level God as an observer, there exists no conflict between religions or between science and religionClassical interactions create quantum probabilities.

....

THE FOURTH WORD:

`` In this excessively enlarged body, the spirit remains what it was, too small to fill it, too feeble to direct

This increased body awaits a supplement of soul

The mechanism demands a mysticism ``

...

THE FIFTH WORD:

We are part of a technologically advanced civilization, we can convert potential energies into motion, and we are aware

That

One little gram of Uranium gives us more than ten tons of coal

That precisely is the problem.

....

THE SIXTH WORD:

``Humanity groans half-crushed under the weight of the advances that it has made

It does not know it makes its own future.

It is for it to make up its mind if it wishes to live``

INTELLIGENCE MUST CONQUER EGO

....

Guess the six words .. Q3

....

TWO NUMBERS (For Q4)

I am sitting in the middle. Einstein sits on my left, and Newton sits on my right. I hand over a paper to Einstein with a number written on it and tell him it is a sum of two numbers. Then I hand over a paper to Newton with a number written on it and I tell him it is the product of the same two numbers. Then there is a conversation between the two of them:

Einstein to Newton: ``I know that you do not know the two numbers``

Newton to Einstein: ``In that case I know the two numbers``

Einstein to Newton: ``In that case even I know which are the two numbers``

Guess the two numbers ... Q4

... ...

CRICKET MATCH (For Q5)

In a 50 overs One Day International Cricket Match between Australia and England, Australia Bats first,and England bats next. At the end of each ten overs the asking rate increases by 0.75 runs per over. England scores 78 in the last 10 overs and lose the match by 16 runs.

What was Australia`s total ... Q5

... ...

FOOTBALL MATCH (For Q6)

A circle inscribed in a Square of side $64^{0.5}$ meters is coloured blue and the balance portion inside the square is coloured red. In a football match between Arsenal wearing blue shirts and Manchester United wearing red shirts the percentage chance of a team winning is equal to the area in sq.m corresponding to its shirt color.

What are the chances of a draw? .. Q6

... ...

DECODE THIS (For Q7)

J EP OPU LOPX IPX UP EFDPEF JU

Decode it ... Q7 `

... ...

PREDICTING THE SENSEX (For Q8)

In November 2008, I wrote an article on ``THE SENSEX AND THE `Y` FACTOR`` about predicting the Sensex for the next 15 years ending April 2024. The article was updated in August 2009 when there was some clarity on the Global situation and there were clear signs of the US recession tapering off.

The full article is available on my website: `www.sixwords.in`

The analysis was based on a factor called `Y-Factor` (where `Y` is simply the Return On Investment, ie the percentage annualized appreciation of the Sensex compounded every 12 months) whose values were first determined according to the movement of the Sensex over the last 30 years based on available data and then predicted for the next 15 years based on the assumption that the broad trend established will continue in the same manner with suitable adjustments after factoring in the recent events as well as the broad parameters that might govern the future trend.

As you can see from the table on page 14 (last but one page) the Sensex Fare value for Today`s date ( September $22^{ND}$ 2023) was predicted - by interpolation - to be about 74600, and the actual value today is about 66000 ie about 88.50 % accurate, and the Fair Value `Y` factor (Index of Appreciation CAGR) for 44.4 years elapsed time since Base Date of April 1079 (Base value 100) was predicted to be 16.06, and the value today is about 15.745 ie. About 98 % accurate.

My next update will be in April 2024 when I will be predicting the Sensex values for the next 15 years ending April 2039.

But an empirical formula for predicting the future Sensex values is attempted here. For this purpose, the past values are first noted according to the movement of the Sensex over the last 45 years based on available data and then predicted for the next 15 years based on the assumption that the broad trend established will continue in the same manner.

So, I found out from Google the Values of the Sensex at close on April 30th of the years in the past - at five-year intervals - beginning with the year 1979, which was the base year of the Sensex and the base value of the Sensex was taken as 100.

These were the values:

APRIL 1979 :100, APRIL 1984: 245, APRIL 1989: 715, APRIL 1994: 3780,

APRIL 1999: 3740, APRIL 2004: 5590, APRIL 2009: 9710, APRIL 2014: 22915, APRIL 2019: 39030

Next, I worked out the Index of Appreciation (CAGR) Factor `Y` corresponding to each of these dates since inception as follows:

`Y` for the period 1979 to 1984 = ((245/100) ^ 1/5 – 1)*100 = 19.63 %

`Y` for the period 1979 to 1989 = ((715/100) ^ 1/10 – 1)*100 = 21.74 %

`Y` for the period 1979 to 1994 = ((3780/100) ^ 1/15 – 1)*100 = 27.40 %

`Y` for the period 1979 to 1999 = ((3740/100) ^ 1/20 – 1)*100 = 19.85 %

`Y` for the period 1979 to 2004 = ((5590/100) ^ 1/25 – 1)*100 = 17.46 %

`Y` for the period 1979 to 2009 = ((9710/100) ^ 1/30– 1)*100 = 16.46 %

`Y` for the period 1979 to 2014 = ((22915/100) ^ 1/35–1)*100 = 16.81%

`Y` for the period 1979 to 2019 = ((39030/100) ^ 1/40 – 1)*100 = 16.09%

Next, we determine the Fair values of this Y Factor corresponding to its true/intrinsic worth in the following manner:

Ignoring the sharp upswings and downswings of the past, it is possible to determine an equation that corresponds to a steady decline in the value of `Y` from 20 in the beginning to 15.745 as on date (22nd September 2023)

Hence, we get: $20*Z^N$ = 15.745, where N = numbers of years elapsed since base date of April 30th, 1979, which is 44.4.

Thus, $20*Z^{44.4}$ = 15.745

Solution of the above equation gives Z = 0.99462

Hence the equation to determine `Y` is: $Y = 20*0.99462^N$, Where N = Number of years Elapsed since base date of April 30th 2079, with base value of Sensex as 100

With N = 45, Y is worked out as = $20*0.99462^{45}$ = 15.689, and the predicted Fare value of Sensex on 30th APRIL 2024 is worked out as = $100*(1 + 15.689/100)^{45}$ = 70490

With N = 46, Y is worked out as = $20*0.99462^{46}$ = 15.604, and the predicted Fare value of Sensex on 30th APRIL 2025 is worked out as = $100*(1 + 15.604/100)^{46}$ = 78840

In the same way by putting N = 50, 55, and 60, `Y` is worked out as 15.272, 14.865, and 14.470 respectively and the predicted value of the Sensex on 30th April 2029, 2034, and 2039 is worked out as about 122000, 204000, 332000 respectively.

THE QUESTION: Given, `Y` which is the index of appreciation (CAGR) is $=24*0.99462\char`\^N$, and N is $=$ Number of years elapsed since the Sensex base date of 30th April 1979, and the base value on the base date as 100.

What is the expected value `V` of the Sensex on 30th April 2030 (V = 100*(1 + Y/100) $\char`\^$ N) ... Q8

... ...

<u>YOU ARE THE POPE (For Q9)</u>

The story goes that Bertrand Russell, in a lecture on logic, mentioned that in the sense of material implication, a false proposition implies any proposition.

A student raised his hand and said: "In that case, given that 1 = 0, prove that you are the Pope."

Russell immediately replied, "Add 1 to both sides of the equation: then we have 2 = 1. The set containing just me and the Pope has 2 members. But 2 = 1, so it has only 1 member; therefore, I am the Pope.

I knew that story ... so one day, when during a meeting, somebody in my office - lets call him `E` - asked me a question:

E: Sir, Who am I?

I took some time (Perhaps three minutes) before I could prepare myself to answer this extraordinarily difficult question.

I told him : I guess you are the Pope.

Complete silence all around .... for several minutes ....

And then I got up gradually and went up to the blackboard and started writing something:

Consider the equation: a = b

Multiply both sides by `a`, we get $a^2 = ab$

Add $a^2 - 2ab$ to each side, we get:

$a^2 + a^2 - 2ab = ab + a^2 - 2ab$

i.e $2a^2 - 2ab = a^2 - ab$

Divide each side by $a^2 - ab$, we get 2 = 1

Silence once again, no body in the room understood what was happening.

E: Sir, It is amazing how you could arrive at the equation 2 = 1, But how is that related to your answer `You are the Pope` to my question ` Who am I `

What was my reply to the question ... Q09

… …

TU TU TU TU TU TU (For Q10)

This happened many years back, in the sixties. I was travelling in a train, it was well past midnight. In a few minutes I was to reach my destination `Jagadhri`. It was not a big station and I knew the train stopped there for a very short time. I was carrying a lot of luggage, and I was nervous wondering if I would be able to take the luggage out in such a short time. I asked the Coach Assistant: For how long does the train stop at Jagadhri ?and he just murmured `Tu tu tu tu tu tu`. I told him I can also sing songs but I am asking a question: For how long does the train stop at

Jagadhri? He looked at me and repeated the words `Tu tu tu tu tu tu`.

So, I didn`t get a proper reply, and became even more nervous.

Anyway, the station arrived, there was heavy rush of passengers alighting the train and passengers entering the coach. I struggled and was just about able to manage to take out my stuff. By the time I took out the last item the train had already started moving and I nearly fell down. The coolie came at last and took control of the baggage. Before moving out of the Platform, I saw a Board Showing the Train Arrival and Departure Times. Against my train it was mentioned:

ARRIVAL: 01.58 am DEPARTURE: 02.02 am

And then I understood what the coach assistant was saying.

Here is the Question: What did the coach assistant say .. For Q10

CRAFTEROOTA (For Q11

Some years back on the Windmill`s Craftworks Annual Day celebrations, I was asked to speak a few words on the occasion. So, I deed the needful and then sang a few lines in tune with a famous Sinatra song `` .......``   ``The Craftwork, The Oota, that remarkable place. The Craftwork, The Oota, and the lovely Terrace. You`ll love it, bet your bottom dollar you`ll lose the blues at Oota, a place that surely someday will be the talk of the town.

The Craftwork, the Terrace, the Oota, I just wanna say, they make things, they do things that just take your breath away.

They have the time, the time of their life, I saw a big crowd they just danced all night, at Craftwork, At Oota, at Terrace .. I congratulate all of you on this Annual Day

Music .. Pau Pau Pau

Repeat last part.

The question: name the sinatra song you are reminded of .. For q11

<u>WELCOME NEW JOINEES AT TOTAL ENVIRONMENT (FOR Q12)</u>

And now `FOR A FEW HONEST WORDS` with those of you who have joined the TE family recently. I `IMAGINE` you will have a good `TIME` with us. `LIFE IS BEAUTIFUL` here in Total Environment. We will provide you with the `WINGS`, but you will have to be `LEARNING TO FLY` yourself and not to let the `RAINDROPS KEEP FALLING ON YOUR HEAD` or get `LOST IN THE GREENS` and remember always that `GREEN IS THE COLOR` of Total Environment. TE will always endeavour to search the `WINDMILLS OF YOUR MIND` and make sure that you always `SHINE ON` and ultimately `REACH FOR THE SKY`. And `AFTER THE RAINS` have brought good cheer, let us all `FREEBIRDS` perform `THE MEADOW DANCE` while the singers sing the `SONGS FROM THE WOODS`, and the artists embroider the `TAPESTRY` of love on this canvas called the `GOOD EARTH`

Guess how many Total Environment Project names in the above para .. Q12

... ...

<u>SCAPRICCIATIELLO BY RENATO CARA SONE (</u>
<u>For Q13)</u>

10 – 8 – 6 – 3 – 1 – 1 – 1 – 3 – 5 – 7 – 8

Any guesses, what is meant by those figures, in the light of this piece of music. I will be immensely – yet pleasantly – surprised if anyone can guess it right.

Well, It was how the song fared in the Binaca Hit Parade heard every Monday from 8.00 pm to 8.30 pm. Eleven weeks on the charts including three weeks at the top.

SILLY .. or maybe .. CRAZY .. but that was me.

The song was also known in English with a single word ``I.........N` of eleven letters with the first letter as `I` and the last letter as `N`.

To say that I am - and always was - a music buff would be a gross understatement.

The year was 1957 ... I was sixteen at the time. Each time I heard this piece of music by the Italian singer `Renato Cara Sone` - there were many others including the instrumental `MISIONERA` by the Trio Los Paraguayos – I could feel the beard on my face. To say that I was enjoying listening to it would again be a gross understatement.

We were living on the 2nd floor in an apartment building called ``NANAK NIVAS` on Breach Candy Road in Mumbai (BOMBAY in those days). Every Monday evening from 8.00 pm to 8.30 pm I was glued to the Radio ( National echo) listening to the Binaca Hit Parade (The top ten in the popularity list), on Radio Ceylon, the DJ was a guy called GREG (I don`t remember the surname). Likewise every Wednesday from 8.00pm to 9.00pm there was this Binaca Geet Mala (The Top sixteen in Hindi Film songs .. Later called Bollywood), where the DJ was the Amin Siani. Greg was also the DJ on various other programs on Radio Ceylon such as `Listener's Choice ` in the evening, and `Wakey Wakey` in the morning. I remember two other Dj`s `Jimmy Barucha` and `Percy Bortholomio`. Greg was very funny and very popular. Frequently he used to tell the time in his own style like: It is twelve minutes past 8.00 pm on this the 14th Day of June, and if you are that lazy, the year is 1957.

I also never missed programmes like the weekly ``Radio Australia Hit Parade`, `The Voice of America Jazz Hour` done by Willis Coniver ( signature tune `Take the A Train`), The Hill billy music ( mixture of blue grass, country & western with the theme song `I`m moving on by Hank Snow) on Radio Ceylon, `A date with you ` on All India Radio Done by Chandrachud who later became chief Justice of India, and many many others. A recall Dean Martin`s `VOLARE ` was number 1 for 10 street weeks in the Radio Australia Hit Parade, and Billy Vaughn`s `RAUNCHY` As number 1 for 8 weeks in `The United States Armed Forces Hit Parade`

Anyway, coming back to the Binaca Hit Parade, One Monday Evening there was no radio in the house, it had been sent for repair and servicing at a nearby radio shop. It was the 5th week of `SCAPRICCIATIELLO` (I.........N) on the hit parade, it was at number 3 in the previous week and I was curious to know where it reached on this Monday, but what to do .. there was no radio. Suddenly at 8.05 pm the hit parade was switched on in a neighbourhood flat on the opp side. I opened our window fully – so that the music was loud and clear- and sat next to it for the entire programme. Dinner was already served but I was not concerned. My dad looked at me and `smiled`. 10 to 4 was done and then came number 3, it was not `I`, if I remember correctly It was a Patty Page number ``ALLEGHENY MOON `, so happy to know that `I` will go up. Then came number 2, it was an instrumental I used to like very much ..`Poor People Of Paris` by Les Baxter dropped down from number 1. I was thrilled `I` would be at number 1. And sure It was. I was listening intently, my dad saw how I was enjoying. When the song was over my dad asked me ``Kuch Samaj Aaya ?`` (Could you understand any thing ?)

No! I had to say .. It was an Italian song .. How could I understand it? .. but the music was too good.

SILLY .. or maybe .. CRAZY .. but that was me. What else can I say.

Any way here is the question: What was the song known as in English.

What was that Eleven - letter word starting with `I` and ending with `N`? .. For Q13

… …

<u>JUST TRIGNOMETRY (For Q14)</u>

WHEELER leaves station `A` at 3.00 pm IST and drives towards North for an hour at 90 kms per hour and reaches station `B`. At `B` he picks up WIGNER and the two of them travel for an hour towards East at 80 kms per hour and reach station `C`. At `C` they have a meeting with a friend `SCHRODINGER` for an hour and fifty minutes. They leave `C` and travel for an hour towards Southwest at a speed of 70.71 Kms per hour and reach station `D`. WHEELER drops

WIGNER at `D` and returns back at `A` precisely at 3.00pm GMT.

At what speed did WHEELER travel from `D` to `A` .. For Q14

... ...

BOLLYWOOD HINDI SONGS SUNG IN ENGLISH (For Q15)

For reasons I do not know why, I had a peculiar habit of translating Bollywood Hindi Songs in English and then singing in English and asking listeners to guess the original Hindi song. There was a time I had worked out the translation for about 40 songs. I have forgotten most of them.

One of them was something like this (in English) :

``I Don`t know where my heart is gone my dear ..

It was lying here somewhere but now its not there ..

Captured by someone whose style is very nice ..

Frightened by someone with her big big eyes..``

Guess the Hindi Song .. for Q15

MAN OF WAR (For Q16)

Man of war was arguably the greatest thoroughbred racehorse of all time. The horse had no idea how to lose a race. He won 20 out of 21 career races. The only race he lost was due to a mix-up at the start. Man of War was not properly lined up and was looking in the opposite direction when the starter called out. By the time he turned around he had lost several lengths. Still he covered lot of ground and was flying near the end to lose the race by a whisker. One more stride and he would have won the race.

The horse who defeated Man of War was called ` ….. ` .

A few days later the word ` ….. ` was coined in the dictionary. It was a five-letter word, the same as the name of the horse who defeated Man of War.

Guess the word … Q16

… … …

"ENLIGHTENMENT OF ...
THE SINGLE MIND

``Seeing the Way is like going into a dark room with a torch; The darkness instantly departs, while the light alone remains. When the Way is attained and the truth is seen, ignorance vanishes, and enlightenment abides forever`` .. Gautam Buddha

To begin with, here is a case study of an enlightenment experience of a man from Kashmir named Pandit Gopi Krishna. Early in life, an essentially well balanced and rational minded agnostic who was as keenly disappointed by the teachings of traditional Hinduism as by the apparent materialism of the scientific outlook, he decided to search on his own without the usual assistance of a guru. Seventeen years of regular meditation went by and disappointingly nothing happened until one fateful day in December of 1937: This time the experience proved to be dramatically different``

At First it was no more than an odd feeling of something moving below the base of the spine, strong enough however to interrupt his meditation. The sensation disappeared and he resumed his meditation. Stronger than before the sensation recurred, became stronger and stronger ...

AND ... *``Suddenly with a roar like that of a waterfall, I felt a stream of liquid light entering my brain through the spinal chord ... The illumination grew brighter and brighter and the roaring louder and louder, I experienced a rocking sensation and felt myself slipping out of my body, entirely enveloped in a halo of light. It is impossible* *to describe the experience accurately. I felt the point of consciousness that was myself growing wider and wider, spreading outward while the body, normally the immediate object of its perception, appeared to have receded into the distance until I became entirely unconscious of it. I was now all consciousness without any outline, without any idea of a corporeal appendage, without any feeling or sensation coming from the senses, immersed in a sea of light, simultaneously conscious and aware of every point, spread out as it were in all directions, without any barrier or material obstruction. I was no longer myself, or to be more accurate, no longer as I knew myself to be, a small point of awareness confined in a body, but instead was a vast circle of consciousness, in which the body was but a point, bathed in light and in a state of exhaltation and happiness impossible to describe.* ` `

Thus started one of the most incredible spiritual odysseys of the twentieth century. But for twelve years after that he could not regain that ecstatic state, and in the words of Amaury de Reincourt (The eye of the Shiva) :

` ` `*He underwent the tortures of the damned and literally descended into the bowels of a Dantean hell. There appeared to be no relief from the seat of kundalini. He began to fear that he was on the verge of insanity. His body wasted away, he could no longer eat and lost all desire to live. So intense was his suffering that he almost made up his mind to commit suicide. But it then occurred to him, being a technically-minded man, that he might have accidentally aroused `kundalini` through the hot solar channel, `pingala`, rather than through the `susumna` within the spine, as he should have. He decided by the supreme effort of the will, to attempt to shift the implacable current to the cool lunar channel ida, on the left side* *to counteract the horrifying fire that was raging inside. Applying all the mental concentration he could muster, he did so.*

The result was astounding:

There was a sound like a nerve thread snapping and instantaneously a silvery streak passed zigzag through the spinal chord, filling my head with a blissful lustre in place of the flame that had been tormenting me ` `.

His first ordeal was over, but there were several others to come and each was overcome.

Reincourt continues to describe in detail about the indomitable willpower of Pandit Gopi Krishna and his growing understanding of the mysterious force that was operating within him, his prolonged research – where his body and physiological apparatus became his own laboratory - into the technicalities of it, as well as his interactions with the greatest Indian saint `Shri Aurobindo` in faraway Pondicherry. Finally after a gap of twelve years PGK experienced that state once again. In his own words this is how he described it :

PGK`s description of his ecstatic state:

` ` *I stopped abruptly, contemplating with awe and amazement, which made the hair on my skin stand on end, a marvelous phenomenon in progress in the depths of my being. I had expanded in an indescribable manner into a titanic personality, conscious from within of an immediate and direct contact with an intensely conscious universe, a wonderful inexpressible immanence all around me. The phenomenal world receded into the background and assumed the appearance of an extremely thin, rapidly melting layer of foam upon a substantial rolling ocean of life, a veil of exceedingly fine vapour before an infinitely large conscious sun, constituting a complete reversal of the relationship between the world and the limited human consciousness. It showed the previously all-dominating cosmos reduced to the state of a transitory appearance and the .... Point of awareness, circumscribed by the body, grown to the spacious dimensions of a mighty universe and the exalted stature of a majestic immanence before which the material cosmos shrank to the subordinate position of an evanescent and illusive appendage.* ` `

Wasn't that something like going through the event horizon of a black hole? In the end it's a question of science, and it is just a question of time before the `mind` in us becomes a `super mind. In few hundred thousand years we have become men from apes, is it not inevitable that in course of time we will become superhuman beings - with super minds - from men? Or perhaps it may not happen, we might self-destruct ourselves, long before that stage.

Pandit Gopi Krishna`s was of course a rare case. But in general, what exactly happens when we get enlightened? Does it remind you of the song ` `There is someone in my head but it's not me` `? From the album ` Dark side of the moon` .... Pink Floyd.

The question may be asked : Why does it happen to only a few of us?

Why are we so privileged?

What have we done to deserve this?

What about the others who don`t get it, whose interactions have been such that they have not met people who can teach them how to attain such enlightenment by meditation or something?

But who can give the answers ? We do not come across people who know the answers. Besides, even if we achieve this state, is it

not selfish to be in that state while there is suffering all around?
Perhaps Sri Ramakrishna – whose disciple Swami Vivekananda
was – might have reached this state of Samadhi sometimes, but
each time he pulled himself together and cried to himself : `` come
down, come down`` in order to continue his mission on earth,
and he pleaded ``Oh mother, let me not attain these delights, let
me remain in my normal state, so that I can be of more use to the
world``

What I am trying to say is that even if we are able to attain this
state, the first thing that will strike us would be `Why am I so
privileged? What did I do to deserve it?

In the words of Fritjof Capra : ``*All schools of Eastern Mysticism
emphasize the basic unity of the universe which is the central feature
of their teachings 'The highest aim is to become aware of the unity
and mutual interaction of all things, to transcend the notion of an
isolated individual self and identify ourselves with the ultimate
reality. The emergence of this awareness - known as enlightenment
- is not only an intellectual act but is an experience which involves
the whole person and is religious in its ultimate nature*``

The question that needs to be answered is : How do we correlate
the spiritual angle with a scientific explanation ? In short, how can
such an ecstatic experience be brought into play? How can it be
objectified and brought within the conspectus of science? How is
this to be done and where is the scientifically minded mystic who
can point the way?

According to ancient Tantra Shastra - Tan means body and
Shasthra means Knowledge (The study of the body) - Texts, the
human body contains 72,000 nadis that channel prana to every
cell. Some are wide and rushing; others are a mere trickle. When

this system flows freely, we are vital and healthy; when it becomes weak or congested, we struggle with poor mental and physical health. But the nadis—like the chakras (psychoenergetic power centers), prana, and other aspects of the subtle body—don't show up under microscopes, medical science has relegated them to the realm of the merely metaphorical. But traditional yogis believe that the subtle body is real, and that understanding it is indeed helpful.

Several books have been written on the subject One of these entitled `AKHANDA` was written by a friend of mine `Dr Shiv Bhushan Sharma` who I met for the first time in 2003 at a Seminar on `Science and Beyond` held at the National Institute of Advanced Studies Bangalore. In `Akhanda` the author tries to explain the threefold manifestation of the soul (Jivatma. Kundalini-Atma and Parmatma) and their respective centres of consciousness. The human being is considered a microscopic representation of the macroscopic universe. The book explains easy and practical approach of Kundalini -Yoga and gives simple scientific methods to evaluate the higher state of universal consciousness.

But how many people have attained that higher state of universal consciousness, and to what degree? That is the question. It is not enough to be well versed with the technology. We need a supreme will power. Pandit Gopi Krishna must be an extremely rare case. He could muster a stupendous will power to achieve his goal, but he did not reveal his secret of how he did it.

And in case if there is no other requirement other than possession of technical knowledge on the subject, then we cannot attribute any spirituality to it. It is in that case to be considered that the mind has been nicely trained as an instrument. Concentration, then is the creation of the instrument: meditation is the use of it. In the

words of Christmas Humphreys, author of `The Buddhist Way`
:

``*Concentration involves contraction of the field of vision, but meditation involves its expansion. In concentration we gain clear vision, in meditation we keep that clear vision but extend it over a larger field and into depths and heights of thought which we have not been able to reach clearly before*``

It is learnt the great Indian philosopher J Krishnamurti, author of the book ``AWAKENING OF INTELLIGENCE` `one of the first modern Indian sages was able to show Westerners how to achieve the state of spiritual serenity and grace called enlightenment. David Bohm The great Scientist and Philosopher and a close friend of Krishnamurthy was asked the question : Was Krishnamurti enlightened? "In some ways, yes," Bohm replied. "His basic thing was to go into thought, to get to the end of it, completely, and thought would become a different kind of consciousness."

But according to Krishnamurthy, meditation need not follow any system, such as systems which advocate sitting in a certain posture, breathing regularly, or practising awareness. All this is utterly mechanical. Another method gives you a certain word and tells you that if you go on repeating it, you will have some extraordinary transcendental experience. According to Krishnamurthy this is just a form of self-hypnosis. Its called Mantra Yoga. By repetition you can induce the mind to be gentle and soft, but it is still a petty, shoddy little mind. Meditation demands an astonishingly alert mind, meditation is the understanding of the totality of life in which every form of fragmentation has ceased. (Refer `Freedom From The Known` by J Krishnamurthy)

Now the question may be asked : What happens at the quantum level? I mean `What kind of wave-function gets collapsed? `

As an observer who attains enlightenment and is part of a certain branch of wave-function must be completely unaware of other branches, because by making a conscious observation and obtaining a particular result .. his `I` becomes that part of the wave-function which completely corresponds to that particular result. (Refer `Many views of the World` .. by E.J. Squires). In other words, the `I` of the observer is part of the wave-function, it cannot be outside the world to observe it.

But there is another kind of Enlightenment that matters more than attaining a higher level of consciousness for some time and then getting back to the normal case, and this other category of enlightenment corresponds to `FLASHES OF UNDERSTANDING` which were not there before.

And I had my fair share of these flashes of understanding which I intend to pass on to the world.

Flashes of understanding such as:

1)   The universe must make sense

2)   There is a Travelling Cosmic Mind that ensures permanent consciousness for all. The TCM is omnipresent in space as well as time. The Cosmic Mind remains forever in the liveable eras of the universe where stars are shining and life and consciousness is flourishing. The TCM does not proceed to the Black Hole and Dark Eras of deep future, as there is no consciousness available there.

3)   There is an Intelligent Field that controls nature. In fact:

a)   The essential material reality is an Intelligent Field

b)   The field obeys the principle of special relativity.

c)   The field links the quantum with the classical.

d)   The intensity of the field at any point is a measure of the information in the field.

e)   The field looks for a biochemistry to give itself consciousness to understand itself.

f)   There isn`t anything else.

4)   There is only one mind in the universe. We all have the same mind. Its just that our consciousness is in the singular

5)   A single mind does all the living and the dying

YES ... ... A SINGLE MIND

ON THE UNITY OF CONSCIOSNESS

`` WE ALL HAVE THE SAME MIND`` are the six words of my first book `` SIX WORDS``

To arrive at a better understanding of the subject, I first provide some Extracts from Erwin Schrödinger's book *What is Life?* and his essay "Mind and Matter``

ERWIN SCHRODINGER`S THOUGHT EXPERIMENT

`` *To declare that of the component cells that go to make us up, each one is an individual self-catered life, is no mere phrase. It is not a mere convenience for descriptive purposes. The cell as a component of the body is not only a visibly demarcated unit life centered on itself. It leads its own life; the cell is a unit life, and our life which, in its turn, is a unitary life consists entirely of the cell-lives.*

`` *Both the pathology of the brain and physiological investigations on sense perception speak unequivocally in favor of a regional separation of the sensorium into domains whose far-reaching independence is amazing because it would lead us to expect to find these regions associated with independent domains of the mind, but they are not. A particularly characteristic instance is the following. If you look at a distant landscape first in the ordinary way with both eyes open, then with the right eye alone, shutting the left, then the other way around, you find no noticeable difference. The psychic visional space is, in all three cases, identically the same. Now, this might very well be due to the fact that from corresponding nerve ends on the retina, the stimulus is transferred to the same center in the brain where "the perception is manufactured"—just as, for example, in my house, the knob at the entrance door and the one in my wife's bedroom activate the same bell, situated above the kitchen door. This would be the easiest explanation, but it is wrong. To understand this, consider a thought experiment on the threshold frequency of flickering, a very interesting experiment:*

Think of a miniature lighthouse set up in the laboratory and giving off a great many flashes per second, say forty or sixty or eighty or one hundred. As you increase the frequency of flashes, the flickering disappears at a definite frequency, depending on the experimental details, and the onlooker whom we are supposed to watch with both eyes in the ordinary way, sees then a continuous light (6 – 004). Let this threshold frequency be sixty per second in the given circumstances. Now, in a second experiment, with nothing else changed, a suitable contraption allows only every second flash to reach the right eye, and every other flash reaches the left eye, so that every eye receives only thirty flashes per second. If the stimuli were conducted to the same physiological center, this should make no difference. If I press the button before my entrance door, say every two seconds, and my wife does the same in her bedroom, but alternately with me, the kitchen bell will ring every second. However, in the second flicker experiment, it is not so. Thirty flashes to the right eye, plus an alternating thirty flashes to the left, are far from sufficient to remove the sensation of flickering; double the frequency is required for that, namely sixty to the right and sixty to the left, if both eyes are open ` `.

What, then, is the conclusion?

` `*It is not spatial conjunction of cerebral mechanism that combines the two reports. It is much as though the right-eye and left-eye images were seen each by one of two observers, and the minds of the two observers were combined to a single mind. It is as though the right-eye and left-eye are elaborated singly and then psychically combined to one. It is as if each eye had a separate sensorium of considerable dignity proper to itself, in which mental processes based on that eye were developed up to even full perceptual levels. This would amount physiologically to a visual subbrain. There would be two such subbrains, one for the right eye and one for the left eye. Contemporaneity of action rather than structural union seems to provide their mental collaboration* ` `

DISCUSSIONS ON THE SUBJECT WITH Dr. MEADOR

Now a question was asked By Dr Clifton K Meador (Author of the bestselling ` `FASCINOMA`) by Email

Dr. Meador's E-Mail of February 17, 2014, to S. K. Sagar

` `Two Visual Cortices: Occipital and Parietal Lobes of the Cerebral Cortex

The occipital visual cortex is at the back of the head. It receives visual images from the retina of the eye. The nerve pathways from the medial retina (closest to the nose) cross back of the eye at the optic chiasm. Thus, the images from the left medial retina crosses to the right occipital lobe, and the images from the right medial retina cross to the left occipital lobe. The left medial retina thus sees objects laterally out in the left visual field; likewise, the right media retina sees objects laterally in the right visual field.

The lateral retinal nerves both go back to the occipital lobe on the same side of the brain. These lateral retinas "see objects" in the central visual fields (central vision). Thus, a destructive lesion to the occipital lobe, say on the right side of the brain, will cause complete blindness to the entire left visual field. The person, so blinded, will "see" nothing to the left of his nose. These lesions to the occipital lobe are quite rare but give an opportunity to study some details of vision.

Now, take a person with a destroyed right occipital lobe, and shine a point of light at him from the blind left visual field. Ask the person to point at the light source, and he will do it with great accuracy with his hand, even though he cannot "see" it consciously.

These light images go to the opposite parietal visual lobe of the brain, located on the top and lateral sides of the brain. These visual cortices of the parietal lobe see form and shape and motions. These are all unconscious visions—not seen in the conscious mind. If we did not have these unconscious visual abilities, all sorts of objects and insects and whatever would fly into our eyes. Our conscious vision is too slow to care for our eyes by averting our heads. The parietal lobes "see" objects, even though the person may be blind from a destroyed occipital lobe.

I am not sure how this all fits with the flicker experiment but thought this background would be helpful in our discussions. `` `

My Reply Dated February 26, 2014, to Dr. Clifton Meador

I have read your notes a number of times with deep interest. Your analysis is perfect, and makes it clear about the physical connection—in time and space—between the right side of the brain and the left visual field, and between the left side of the brain and the right visual field.

Now, how does all this fit with the flicker experiment? Consider Charles Sherrington's passage "It is not the spatial conjunction of cerebral mechanism, which combines the two reports; it is as much as though the right-eye and the left-eye images were seen each by one of two observers, and the minds of the two observers were combined to a single mind, and so forth."

What I believe Sherrington is trying to say is that, even though there may be a physical connection in time and space, it is not this spatial conjunction (presumably this connection in time and space) which combines the two reports. I think we have to see it in the light of the nature of the thought experiment, which provides us with an initial condition that there is a suitable contraption which allows only every second flash to reach the right eye, and every other flash to reach the left eye, so that every eye receives only thirty flashes per second, which is not enough to remove the sensation of flickering. So, if there are two observers (each with both eyes open) and the same contraption is used, which allows only every second flash to reach one of them, and every other flash to reach the other one, each observer will see only thirty flashes per second, and the flickering sensation persists. Now, if the contraption is removed, and there are sixty flashes per second, the flickering sensation goes, and each observer sees a continuous light even if he chooses to close one of the two eyes. This means that there is a *mind* that makes the connection between the eyes of an observer, and it is the same mind that makes the connection between the two observers, except that, physically, this connection between two observers does not materialize, for the simple reason that the consciousness of each is in the singular.

Thus, in each of the two examples (that is, your analysis and the flickering light thought experiment), the common denominator is that there is an intelligent field, which we call the "mind," which designs and builds this physical connection. As Sherrington says, "Contemporaneity of action rather than structural union seems to provide their mental collaboration."

My favorite Sherrington passage is the one starting with the following line: "Are there thus quasi-independent subbrains based on the several modalities of sense?" It ends with this line: "Matter and energy seem granular in structure, and so does *life*, but not so the *mind*."

In between these lines is the following passage: "the single nerve cell is never a miniature brain." Now, it is true that a single nerve cell inside us is never a miniature brain; it is just one out of trillions and trillions of them, each one, though aware in the singular, is just an insignificant part of us, and we, in turn, are conscious in the singular. However, our consciousness is the sum total of the awareness of these individual cells; when we consider a single-celled independent creature such as a paramecium, it looks as if this single cell is indeed a miniature brain, for it can swim toward food, retreat from danger, and negotiate obstacles. This is curious, but it strengthens my viewpoint about the ever-present intelligent field and how a mind, when it encounters a certain biochemistry, acquires a component of this intelligent field and gets awareness.

With all this, I have a long way to go to develop a full understanding of the subject. It is extremely complex, but it is at the heart of my book *Six Words*. It is my good fortune that you are taking so much interest in my book. I stand to gain immensely from these interactions with you.

FURTHER CONVERSATIONS WITH Dr MEADOR:

From Dr. Clifton Meador to S. K. Sagar (February 28, 2014)

``Bruce Lipton in his book *Biology of Belief* makes the case that each cell is a mind. I'm not sure I agree with all that he says, but I thought it would interest you to see the book, which you may already know. He posits that the cell membrane is the brain of each cell—some far-out thinking for you to consider.``

From S. K. Sagar to Dr. Clifton Meador (March 15, 2014)

Regarding Bruce Lipton, I started watching the video on "Belief in Biology." It was very interesting; it's an hour-long video, and he's talking too fast, so I decided that I'll buy the book. In the meantime, on the general concept that "each cell is a mind," I have this to say:

If this were so, it would mean the mind is something physical, something made out of matter such as atoms and molecules. However, the world of matter must comply with the laws of physics, which on the macroscopic level (ignoring quantum effects) are deterministic and mechanical and, thus, incompatible with free will. The mind, I believe, is not the cell itself but is something within the cell, and it gives it intelligence via its intelligent field. It gives the paramecium (a single-celled creature) the intelligence to swim toward food, retreat from danger, and negotiate obstacles, thus disobeying the laws of physics—not just Newton's laws of motion but also the second law of thermodynamics, whereas a dust particle, which is outside the living organism, can do nothing of the sort.

Descartes describes that "something" within the cell:

It is not a physical substance, but a tenuous, elusive ethereal sort of substance, the stuff that thoughts are made of, free and independent of ordinary ponderous matter. The mind is neither perceptible by the senses, nor extended in space. It is intelligent and purposive, and its essential characteristic is thought, or rather, consciousness, or maybe just awareness in some cases. Though the human body is an engine, it is not quite an ordinary engine, since some of its workings are governed by another engine inside it— this interior governor-engine being one of a very special sort. It is invisible, it is inaudible, and it cannot be taken to bits, and the laws it obeys are not those known to ordinary engineers. Perhaps it's a "ghost in the machine."

There is, however, another alternative that does not require a soul, but in this alternative, in which there is no soul, the mind must play a double role. As Schrödinger says: (`What Is Life and Mind And Matter` `)

` `*The conscious mind plays a curious double role, On the one hand it is the stage, and the only stage on which this whole world-process takes place. On the other hand, we gather the impression, maybe the deceptive impression, that within this world-bustle, the conscious mind is tied up with certain very particular organs (brains), which while doubtless the most interesting contraption in animal and plant physiology, are yet not unique, not sui generis; for, like so many others, they serve after all only to maintain the lives of their owners, and it is only to this that they owe their having been elaborated in the process of speciation by natural selection* ` `.

The other difficulty with that concept is that it leads to trillions and trillions of minds in the universe, with each one dying when the cell dies. This does not agree to Schrödinger's concept of the oneness of mind, of which I am an outright supporter. The show that is going on in the universe obviously acquires a meaning only with regard to the mind that contemplates it. It cannot be that the mind was produced by that very display that it is now watching and would pass away when the sun finally cools down and the Earth has been turned into a desert of ice and snow.

So, at this stage, I am not particularly inclined to agree with the concept that each cell is a mind, but I have an open mind. Indeed, this is an extraordinarily complex subject, but is to me the most interesting subject, and it is at the centre stage of my book *Six Words*, and I am grateful to you for generating a detailed, in-depth discussion on it.

....

BUILDING BRIDGES

This is a beautiful peace written by Dr. Clifton Meador.

It tells the story of the marriage of his daughter MKMeador and my son Nikhil Sagar. about the bridge that was built to unite the two families `Sagars` and `Meadors`. To some extent it was a bridge to unite `Bangalore` with `Nashville`.. and to an even more limited extent a bridge to unite `India with US`.

Passage to India ... Clifton Meador

Wedding of daughter Mary Kathleen Meador (aka MK) to Nikhil Sagar.

On the afternoon of April 18, 2014, members of my family gathered at the Lufthansa gate in O'Hare airport in Chicago. We would soon board the flight to Frankfurt, Germany and from there to Bangalore, India. Twenty-three hours in flight.

Daughter Buffy (Driskill) flew in from Norfolk, Virginia. Son Clif and his wife Mary Neal arrived from downtown Chicago, where they live. Daughter Ann (Shayne) and I flew up from Nashville. Son Graham came over from his flight from Nashville to Midway. By 6:30 we were airborne on the largest airplane, a 747, two story office building capable of flight. Amazing to me.

I splurged and booked a business class seat, complete with a real bed, TV set, earphones, and a menu of food and wine. I was able to sleep on both legs of the flight.

We arrived in Bangalore at 1:00 AM to find Nikhil and MK at the airport with transport cars and full time drivers. We stayed at the Taj West End hotel. This grand hotel was originally from the era of British colonization, dating back over 125 years. There were 22 acres of beautiful gardens, pools, a spa, and many palm and other tropical trees and plants and animals. The bird sounds were intense each morning.

Bangalore is in the tropics at around 12 degrees latitude (Miami is 25 degrees) so the weather is quite mild, sometimes hot. We had temperatures in the 80s and cooling breezes the whole time we were there, missing the monsoons which come in May and June.

After a good night's sleep, we gathered for breakfast and a feast of Indian dishes which continued the rest of our stay.

Each night we had a different agenda. We met the next night at the Sagar's home apartment to meet Nikhil's parents S. K. Sagar and his wife Bharati. S.K. has just published a book "Six Words," a tour de force covering astrophysics, atomic physics, evolution, cosmology, and much of philosophy. He is attempting to reconcile existing scientific knowledge with religion. I recommend the book highly and I spent several hours discussing the subjects with him. He is the structural engineer for his son's architectural firm.

Bharati, Nikhil's mother, is an established artist and portrait painter. The walls of their apartment display her beautiful work. Both parents are kind and generous people, and MK is so lucky to have married into such fine family.

Nikhil's brother is Kamal, and his wife is Shibanee. Both are remarkable people. They have 3 sons, a teenager and twins who are 7, I think. Kamal is an architect and developer. We visited his most recent development which consists of 7 tall buildings in a 20 acre setting. He designs and constructs condos and apartments in these building. These can be 1 or 2 or 3 story condos, reaching all the way to the roof if desired. Kamal's condo reaches to the roof where he has a garden with a mowed lawn around his deck, looking out and across Bangalore. His music listening studio is fabulous as is his collection of recordings. Shibanee is also an artist. She created large animal forms made entirely of flowers for one of the evening events.

One night we exchanged gifts with the Sagars. I took Goo-goo clusters. S.K. is a country music fan and he loved the notion of Goo standing for Grand Ole Opry!

The wedding was on Thursday April 24th. It started at 7:00 AM, actually more like 8:00. The ceremony took place in the gardens of the hotel. Clif, Graham and I got formal coats and outfits for the wedding and added turbans that morning. The Sagar family, with Nikhil mounted on a decorated horse, came walking up the road where our family met them. SK and I exchanged garlands of flowers, indicating a union of the two families.

MK is covered in beautiful silks and flowers. Her hands and arms are covered in Henna decorations. Overall, she is a beautiful bride. Nikhil is decked out in a white long coat.

We sit on an elevated platform with long silk ribbons and covering the top and sides. On one side, Clif, Mary Neal and I sit. In the center Nikhil and MK sit and on the opposite side are 2 Hindu priests, called pundits. The rites are VEDIC and usually in the ancient Sanskrit language. Bharati had the ceremony translated into English and I have inserted the steps here

Marriage Ceremony as per VEDIC Rites

Marriage ceremony as per Vedic rites binds the Bride and Bridegroom in marital bond of love and respect for each other and also lay down a code of conduct towards their family and the society. Every phase of the ceremony has its own symbolic meaning & spiritual significance.

The Ceremonial Procedure:

The auspicious Marriage Ceremony begins with the Bride and Bridegroom exchanging garlands symbolizing their acceptance of each other as their life partner. After welcoming the Bridegroom the Bride requests him to be seated. She offers him water to sprinkle on his feet and face to refresh him. She then offers water to him to drink.

Madhuparaka Ritual:

The Bride offers the Madhu parka{A combination of honey, ghee & curd} to the Bridegroom . Before taking a part of this Madhu parka, the Bridegroom sprinkles it in all directions, expressing his hospitality to the all guests participating the ceremony. It also expresses the determination of the Bride to turn any sourness in their relationship into sweetness.

Kanya Pratigrahan/Kanya Daan:

Kanya Daan is an emotional moment when the Bride is given away in marriage to the Bridegroom by her parents. The groom avers that he is happily accepting her hand.

The formal pledge:

The Bride and the Groom addresses the guests witnessing the ceremony that their hearts will always be united just as water from two separate tapes or glasses when poured in one bowl becomes one and can not be separated, similarly their hearts will be united and never to be differentiated.

Agnihotra:

Now the time comes for Agnihotra /Havan before the Panigrahan Sansakar. Havan is performed by chanting Holy Ved Mantras ,oblations are made to fire and prayers to The Almighty God to bless the couple.

Panigrahan Sanskar:

The Bridegroom holds the Brides right hand for trust, co-operation and promises to keep her happy by giving her marital status of a wife in his life.

The Bride and Groom go round the fire once and stand in their respective places. The Groom chants a mantra defining their relationship in a poetic language.

Shilarohan:

The Bride places her right foot on piece of rocky stone , symbolizes for steadfastness events may occur in married life.

Laja Home:

The Bride offers puffed rise (Laja) as oblation to fire with special prayers for the longevity and well being of the Groom. It is followed by three parikrama round the fire in which the Bride holds the Groom"s right hand. In the fourth round the Groom holds the Brides hand and perform the parikrama, symbolizes taking the Bride to his own house.

Saptapadi:

The seven steps taken together by Bride and the Groom symbolizes the seven vows taken by them to make their marriage sanctified and success.

Ist. Step is for ISHA---It is for the fulfillment of the material needs of the family and for Prosperity.

2nd. Step is for URJA---It is for the development of physical, mental and spiritual strength.

3rd step is for RAYASPOSHA---To acquire wealth by pure and righteous means and to spend it wisely .

4th.step is MAYOBHAV---To develop harmonious relationship and be happy.

5th. Step is PRAJA---(progeny) To excel in raising strong & virtuous children.

6th. Step is for RITU---For togetherness in all times and compatibility.

7th. Step is for SAKHA---means friendship-to be dependable and faithful to each other and life long companionship.

After taking the seven steps the Bride and Groom go round the fire together and water is Sprinkled on couple"s head by Grooms mother.

Suryadarshan:

Prayers to SUN , symbolizes to get energy from the Sun.

Hirdaya Sparsh:

Bride and Groom placing their right hand on each others hearts avowing to be loyal to each other.

Sindoor Daan:

Sindoor (vermillion powder) is applied by the Groom in the parting line of Brides hair to signify their sanctified marital relationship, and feed sweets to each other symbolizing their sweet relationship.

Dhruva Darshan:

For strong, steady relationship and togetherness.

Blessings to newly wedded couple:

Showering the flower petals on the newly wedded couple by the Priest, family members and the guests for their happy, healthy and prosperous life.

The entire ceremony was moving. The pundit read the words. Periodically the second pundit would chant long phrases in what I assume was Sanskrit.

There was one funny part that brought laughs from the audience, who were all seated in chairs behind a field of tables with white table cloths. The pundit, speaking to the bride and groom said and here I am taking liberties in my memory. He said something like this. "From time to time, you, the groom must realize that your bride may become petulant, even sharp tongued. You must always respond with sweetness. Always remain sweet. And to remind you, take of this Madhu parka a combination of honey, ghee, and curd. Its sweetness will always remind you of your need to remain sweet." Nikhil then ate a bowl of the sweet curd. It was a particularly touching part of the ceremony and the crowd chuckled when the pundit said those words.

Clif, my son, passed the rings to the couple. My role was to pour water on their joined hands, indicating a union not to be separated. I was impressed with how much the ceremony was a set of instructions to the couple on the essentials of a good and lasting marriage. I thoroughly enjoyed being there.

The entire trip was such a treat for me. Only Aubrey and Rebecca could not make it. The others spent good time visiting. Ann and Mary Neal with Clif took off to see southern India. Graham rented a motorcycle and took off to who knows where. Buffy and I got on a plane to Frankfurt Saturday April 26 to Frankfurt and then onto Chicago for me; New York and then home for Buffy to Norfolk. Nikhil and MK would return to Chicago a few days later.

I did not cover the incredible traffic of Bangalore or so much more of this fascinating country and city, half way around the globe. I have many good memories of MK's new family. She is lucky and blessed to have found Nikhil and his family.

· · · · · · · ·

ROCK ISLAND LINE

In a parallel universe on the planet Earth ., a time will come when the world will see the light of day … about what happened on 9/11/2001.

Perhaps George Bush will deliver the truth just in time before exit time.

Remember the song `Rock Island Line` and the lyrics:

`` And so the train moves through .. and when the train moves through, it takes up a little bit of steam, and a little bit of speed, and when the driver... he thinks he is safely on the other side .. he shouts to the man (at the toll gate) and he says: `I fooled you, I fooled you, I got pig ions, I got pig ions, I got aaallll pig ions` ``

In the same way the `Bush` when he thinks he is safely near the other side of life .. might say: `I fooled you, I fooled you, I brought the building down, I brought the building down, I brought aaallll buildings down` .. and then wait for the big drama to unfold.

And then some guys from the Deep state will try to control the situation and prevent the situation from getting out of hand. And they will say: He is just talking rubbish .. just to log the limelight. We all know, it was the Al Quida that brought the buildings down. We all know that a certain Class 757 plane expertly piloted by a certain `T` hit the Tower at a certain speed and continued to travel at the same speed - piercing the strong structural framework of the building till it was completely inside, and all this happened not once, but twice.

A certain `SKS` will then look for Newton`s Principia and see if somewhere in the epilogue of that book Newton might have said something to this effect: `I fooled you, I fooled you, I told the wrong law, I told the wrong law, I told aaallll the wrong laws`

For further reading refer my book ``IS THIS OUR FINAL MILLENNIUM``

Anyway, here is my version - in my hoarse voice - of the skiffle ``ROCK ISLAND LINE``

(Visit my page `SKSAGAR100` on You tube)

… … …

BIOPSY OF THE DEEP STATE CANCER

On 28th October 2022, I was in Fortis Hospital, Bangalore for Prostate Biopsy (TRANSRECTAL ULTRASOUND GUIDED BIOPSY)

Six Cores each of tissue were obtained from bilateral halves of prostate .. total 12 cores.

I was informed later that I tolerated the procedure well .. that does not mean It was not painful. Imagine 12 times drilling holes in the prostate and removing core samples to be tested later for cancer.

Somewhat similar to concrete cores used for testing of actual properties of concrete in existing structures such as strength, permeability, chemical analysis, carbonation etc. Consider for example an RCC Column a portion of which is of Suspect strength based on earlier cube tests. Indirect testing such as Ultra sound Pulse Velocity tests also revealed suspect quality so it becomes necessary to drill holes and take out core samples and test them to enable a more direct assessment on strength and quality. So, if the result shows low strength, the column will need to be dismantled, and if it is somewhat low or if it so happens that the floor beams/slabs supported by such columns are already cast, then the column needs to be retrofitted by such measures as Jacketing the column etc.

In the same way if the Biopsy reveals malignant cancer in the prostate and depending on the extent of cancer it needs to be decided if the Prostate needs to be removed by Surgery, or retrofitted by some other treatment such as Radiation Therapy or Proto Beam Therapy

Using the same analogy .. I think it is extremely necessary to drill holes and take out large number of core samples .. from different locations (in this case `different years`) of American Foreign Policy to check for suspect quality and consequent damage caused in terms of Millions of lives lost etc.

This is precisely what I have done in my book ``IS THIS OUR FINAL MILLENNIUM? (Second Edition), and found out to my horror that nearly every sample failed, and the AFP is comprehensively saturated with a Cancer known as ``DEEP STATE PHENOMENA``. The damage is so extensive that there appears to be no other solution except to remove the cancer ie Dismantle the Deep State Phenomena. But the trouble is .. A substantial part of the `DEEP STATE` Cancer has come out of America and spread to various other countries such as China and Russia. Just as - as many doctors believe - the cancer inside the Prostate can come out of the prostate and travel through the blood stream and infect other parts of the body such as Bones etc, in the same way a portion of `DSP` has come out of AFP` and travelled to China and Impacted/infected the CFP, as is evident from the Coronavirus Epidemic transmitted to the world by China, especially to America. In case of Russia the impact/infection on the `RFP` has happened in such a Chaotic Manner that the RFP is struggling to deal with it and has no idea what it should do, and in such a situation it can do something silly such as resort to a Nuclear Attack on the US.

So, the `DSP` needs to be removed.

But there is a problem .. And the problem is : The DEEP STATE CANCER is so widespread throughout the world – in particular the SUPER POWER WORLD of US, RUSSIA ,

AND CHINA, that it is literally impossible to Remove it simultaneously from the `AFP`, `CFP` , and `RFP` etc.

Which means the only solution is `RETROFITTING`

This is precisely what I have endeavoured to do in my book ``IS THIS OUR FINAL MILLENNIUM? ` Second edition, by suggesting Retrofitting in the form of TRANSFORMATION of the world leaders .. specially those of US, RUSSIA, and CHINA, by transmitting GOOD SENSE to them which can make them to do some `SELF REALISATION` and some `RETROSPECTION`

What could hopefully achieve this critical transformation is bringing all the critical personalities involved in the major geopolitical, religious and commercial conflicts to the table of dialogue with open mind to end war, shed personal ego and nationalism and solve all the ongoing conflicts with the single criterion of peace and just and best interests of all people everywhere and forget and forgive all the rest that is divisive.

Who should be assigned this onerous task of persuading the transformation of all critical personalities? My Answer: Men and women reputed for their pursuit of peace, justice and continued all round progress of humanity.

As suggested by my friend Mr. AK Chandrashekhar, The book needs to reach pronto all UN members and its Secretary General, the hands of both the critical personalities to be transformed and the credible persons of eminence who can effect such a transformation soon enough to save humanity from extinction.

....

A QUESTION OF SCIENCE

For a change I do not involve myself in what is right and what is wrong.

For a change I do not profess to have any understanding of `The Deep State Phenomenon`. and whether the real people who manipulate and control their government`s policy to stage manage conflicts and wars are behind the visible politicians and they control the politicians, and those people are happy and successful at what they're doing and it is difficult to change them.

For a change I do not profess to understand whether these deep states as well as the fused states of the world continue to play havoc with human lives just to ensure that the business is flourishing in certain categories of industries that cannot survive in peaceful times.

For a change I do not profess to understand if it is an aspect of design by the `UNINTELLIGENT FIELD` that 1) conflicts and wars must go on and on, so that the military industries of the world keep flourishing, never mind if millions of innocent civilians including soldiers in the army are dying and the common people keep paying the taxes to support the huge Military Budgets, that 2) Pandemics and Bio warfare must go on and on so that the Pharma Industries of the world keep flourishing, never mind if millions of people are dying including Doctors and Nurses and other Health Workers, that 3) if these industries are flourishing and there are conflicts and wars going on and on, and there is lots of news to tell, it is likely that the Media Industry will also keep flourishing.

For a change I do not profess to understand if our universe, even if it is meaningfully informed, is still one in which great evil can exist alongside good and that the real world has to be a mixture of harmony and nuance.

All I want to do is ask a question of science, answer it myself and then challenge any scientist in the world to come forward and prove me wrong.

MY QUESTION: Given the structural framework of the two WTC towers, is it possible for the towers to fall precisely in the manner in which they actually fell (as seen in the video recordings available) by any other means other than by a scientifically and meticulously planned controlled demolition ?

Each tower was 64m square, standing 411m above street level, and 21m below grade. The Structural design comprised of a lightweight "perimeter tube" consisting of 244 exterior columns of 36 cm square steel box section spaced 1m c/c. Inside this outer tube there was a massive 27 m × 40 m core, made up of a framework of heavy columns, fully braced, which was designed to support the weight of the tower, and of course the interconnecting beams. Steel beams 800 mm deep connected the core to the perimeter at each story. Concrete slabs were poured over these beams to form the floors.

MY ANSWER TO THE ABOVE QUESTION: An emphatic NO. It is definitely not possible without violating the established principles of Structural Engineering which are in turn based on the established Laws of Science.

NOW, I hereby challenge any scientist or engineer on the planet to come forward and prove me wrong.

CHEERS

A QUESTION OF ENGINEERING

Now, a question may be asked: Such a meticulous plan involving drilling of holes at hundreds (perhaps thousands) of locations (precisely worked out) and planting explosives etc required to ensure that the structure must collapse inwards and onto itself, must have required several days of planning involving many many meetings. While, an implosion like this requires explosives and detonators, it's not simply just setting the charge and punching the detonator to trigger a blast. It also relies on gravity for the building to collapse in a pre-determined pattern. For it to happen, engineers will have to tactfully remove the supporting structure of a building at a certain point that brings it down upon itself. To do so, explosives are required to be drilled into key components of the building that support the structure. After the detail design was carried out by the Structural Engineers, several drawings – perhaps in dozens for each tower - would have been required to be released for construction/demolition. Where were the meetings held and where are the documents/drawings available. The job must have been entrusted to contractors, one for each tower, again involving several meetings and preparations and approvals of contract documents. Where were these meetings held and where are the documents/ Records available.

Several high-level meetings would also have been necessary, between members of the Deep State and the leaders of the terrorist organizations, to prepare the masterplan for the event and obtain the President's approval. It might be too far-fetched, but I think there must have been some meetings between the President and OBD.

Where were all these meetings held and where are the documents/Records available or hidden?

THAT IS THE QUESTION

MY ANSWER TO THE QUESTION: I guess the planning was done in the third WTC Tower. All the Documents, Drawings, Records, Minutes Of meetings, etc etc must have been kept / hidden somewhere in the building. But there was a huge risk involved of the documents being discovered in the future, so it was an absolute necessity to bring down the third WTC Tower. And we all know that the Third WTC tower was also brought down by controlled demolition on the same day and in the same manner as the other two towers were brought down.

And there were no planes that struck the third tower.

A QUESTION OF RATIONAL THINKING

So, we are living in a false world, and the common man in the street is unaware of it.

But the truth cannot be hidden forever. Dissimulation is of no avail. Dissimulation is to no purpose before so great a judge. Falsehood puts on a mask. Nothing is hidden under the sun. if we find from our own experience that something is a fact and it contradicts what some authority has written down, then we must abandon the authority and base our reasoning on our own findings.

And there are enormous and very serious implications if the truth is hidden for long, such as : The Deep State of America will continue to play havoc with human lives across the world - America Included – Conflicts and wars will continue and even escalate - ultimately turn into nuclear wars.

Yet there are also serious implications if the truth is revealed .. particularly if it is done in a dramatic manner, such as : It can cause a Knee Jerk Reaction such as a sudden nuclear attack by the US itself, spearheaded by its Deep State, with an intent to tell the world `WE ARE WHAT WE ARE` or in the words of Noam Chomsky ``WHAT WE SAY GOES``. Surely, such an attack will lead to a massive chain reaction of nuclear attacks followed by a nuclear winter of some kind.

But the truth has to be revealed .. perhaps with a SOFT approach with an understanding that a nation is not to be blamed, as a nation is just a geographical area, it's the Deep State Phenomena that needs to be taken care of.

This is a complex situation.,

How does the `INTELLIGENT FIELD` resolve the problem?

The trouble is … The intelligent Field is unaware of what is right and what is wrong, It can only process the Information that is transmitted to it

Nature is the source of all true knowledge, she has her own logic, her own laws, she has no effect without cause nor invention without necessity. Our job here is to determine and show the sequence of events that turn out to be such causes that lead unto such effects which leads towards peace in the world.

So, we select the most appropriate information from the field such as the wisdom from the past great scientists, and philosophers of the past and transmit the same to the current world leaders who matter, and whose transformation – from destructive to constructive - is essential for world peace.

What is my role here?

And why should I have to play a role here?

That is the question.

Perhaps the writing of this book is the `Effect` of a certain `Cause` such that there was no way I could have avoided it.

So, It is not that I wrote this book .. It is that it was written by me, consequent on the interactions of the world on me. I even do not know, if this, the writing of the book is attributed mainly to the revolving doors of chance, or to the existence of a certain "intelligent field", that has created probabilities that books of this kind should be written to prevent the human civilization from getting extinct. There is also an "Entropy field ", which is continuously increasing disorder and chaos and leading us towards self-destruction. We do not need to solve complex differential equations to understand and realize how far the `intelligent field` is trailing behind the `entropy field`. Will the former be able to overtake the latter before it reaches its winning post of destruction. It is entirely left to us human beings. No God of any kind will come to our rescue.

… … …

TIME FORM

EVALUATION OF TIMEFORM RATINGS

In this article a procedure has been developed to determine the Time form rating of a thoroughbred racehorse based on its performances on the track. At first, let me explain some of the terms used in racing vocabulary by answering questions such as:

What is meant by the term LENGTH `L` and what is the time required to cover 1 L:

Length corresponds simply to the length of a horse. There are no statistical surveys carried out to show what the average length of a horse is or what the average height of a human being is. According to my knowledge the average length of a thoroughbred horse is about 2.70 Meters (8.85 Feet).

What is the time required to cover one length particularly at the finishing stage? This varies a great deal, in a closely fought 1000 m sprint in the highest class on an excellent track in good going this may take as little as 0.14 seconds. Tiring horses in a long-distance race, back markers (no hopers) may take even more than 0.28 secs. On an average in a close finish a horse may take about 0.165 seconds to cover one length.

What is the effect of increasing the WEIGHT on a horse?

Increasing the weight on a horse by 1.0 kilo will render a horse slower by $0.0001D$ seconds, where D = distance of the race. Thus in a 1400 Meters race a horse will take 0.14 seconds longer than what it would have taken if he had carried 1.0 kg less weight. The effect per kg increase in the weight is even more pronounced if the initial weight is in the range 60 to 70kgs and substantially more pronounced if the initial weight is greater than 70 kgs. Further some horses are more susceptible to weight increase and others (called weight carriers) are less susceptible.

What is `RATING` & How to determine the RATING of a horse:

Rating of a horse signifies its caliber as a race horse. Standard rating system adopted In U.K. and many other countries is the TIMEFORM rating system and the rating points given are in terms of pounds by weight. Thus, a horse `x` rated 5 points higher than another horse `y` is expected to run as fast as `y` even if carrying 5 lbs higher weight provided other parameters are equal. Highest Time form rating given to date is 145 to the champion horse `Secretariat 'Though `Frankel` was rated at 147 by the British Experts, I am not very sure it was correctly evaluated.

I have developed a formula to determine the Time form rating of a horse based on speed records for various distances giving due cognizance to the weights carried and to the fact that some of the records might have been achieved on an unusually fast track.

First I give below the World record timings for various distances varying from 5 Furlongs (1006m) to 2 miles (3218m):

1006m: PROCREATE - 7 yrs, Weight 120 lbs, Track Hollywood Park, 1995- 53.79 secs(unusually fast dirt track)-on turf the record is 55.20secs held by CHINOOK PASS

1207m: G MALLEAH - 4 yrs, 120 lbs Turf Paradise – 1995 – 1min.6.6 secs

1408m: SOARING FREE – 5 yrs, 126 lbs, Woodbine – 2004 – 1 min 19.38 secs

1610 m : MR.LIGHT – 6 yrs, 118 lbs, Gulfstream Park—2005 -1 min.31.41 secs

1811m: KOSTROMA- 5 yrs , 117 lbs, Santa Anita Park – 1991 -1min.43.92 secs

2012m: DOUBLE DISCOUNT- 4yrs,116 lbs,SantaAnitaPark-1977-1m57.4 secs

2213m: WITH APPROVAL --4 yrs, 118 lbs, Belmont Park – 1990 -2m. 10.20 secs

2414m: HAWKSTER – 3 yrs- 121 lbs, Santa Anita Park – 1989 – 2m 22.80 secs

2816m: PAPER JUNCTION- 4 yrs 123 lbs, Lincoln State Fair 1985 2m.50.4 secs.

3218m: PETRONE – 5 yrs, 124 lbs, Hollywood Park- 1969 – 3m-18secs

Next, by careful scrutiny of the above record timings (ignoring some freak runs) and on the assumption that as the distance increases the timing for each 200m increases from about 11.5 secs at the 1000m stage to about 14.0 secs at the 3000m stage (This range is for top rated horses increasing to about 13 secs & 16 secs respectively for zero rated horses) I determine the yardstick for a 150 points rating on the Timeform scale being the time taken `T1` by a horse rated 150 points to cover various distances carrying 120 lbs(54.5kgs) on a perfect and fast track, with perfect going, with perfect distance suitability with no false rails etc etc as follows:

Distance: 1000m Time T1: 54.7 secs

Distance: 1200m Time T1: 66.4secs

Distance: 1400m Time T1: 78.4 secs

Distance: 1600m Time T1: 90.7 secs

Distance: 1800m Time T1: 103.2secs

Distance: 2000m Time T1: 115.9 secs

Distance: 2200 m Time T1: 128.8 secs

Distance: 2400m Time T1: 142.0secs

Distance: 2800m Time T1: 169.0 secs

Distance: 3200m Time T1: 196.7secs

Next, I give below the formula for determination of the Time form rating of a horse based on a particular performance

Let T = Time in secs taken by the horse.

Let D = Distance traveled in Meters.

Let W = Weight carried by the horse in kgs

Then its Time form rating = $2.22W + 30 - 22222 (T - T1)/D$

Some examples based on the assumption that perfect conditions materialized are given below:

MYSTICAL`S performance in the Indian St. Leger of March 2006, considered as the best performance seen on an Indian racetrack.

Distance `D` = 2800m, Weight `W` =57.0kgs, Time `T` =2min 54.06 secs =174.06secs Hence TF Rating = 2.22 x 57.0 + 30 – 22222 (174.06 –169.4)/2800 = 119.5

SPECTACULAR BID`S performance at Santa Anita in 1980

Distance `D` = 2012m, `W` = 56.75 kgs (126 lbs), Time `T`=Im 57.8s=117.8 secs, Corresponding `T1` for 2012 m = 115.9 x 2012/2000 = 116.6 TF Rating = 126 + 30 – 22222(117.8 – 116.6)/2012 =142.7

SECRETARIATE`S performance at Belmont Park in 1973 considered by many as the best performance seen on any race track in the world. Distance `D` = 2414m, `W` =56.75 Kgs(126 lbs) Time `T` = 2min24 s= 144 secs Corresponding `T1` for 2414m = 142.2 x 2414/2400 =143.0secs TF Rating = 126 + 30 – 22222 (144 – 143.0)/2414 = 146.8

ADJUSTMENTS IN TIMINGS

The basic requirement for a system to yield `absolute` values of the Time form rating of horses is that either the track conditions must be perfect, or they must be perfectly understood. For comparative values it might be enough to know the Short-Term Track Index `Ts` which is a measure of the Penetrometer reading values that determine the day-to-day variations in the track conditions caused by rain and storms. But for `Absolute` values it is imperative to know the Long-term Track Index `Ti`, Defined as the distance travelled by a horse rated 150 on the TF scale carrying 120lbs weight in a time of one minute on that track when that track is at its fastest with least Penetrometer reading value)) of the course for all distances. Accordingly, an adjustment in timing is required after taking into consideration both the long term as well as the Short term track indices.

1) DUE TO LONG TERM TRACK INDEX

The procedure for arriving at the Time form rating as outlined above will give accurate results only if perfect track conditions materialize and the horse runs to its full potential. However, there is a wide variation in the track characteristics of tracks worldwide. This is attributed to several factors such as ground characteristics (nature of ground, dirt or turf, thickness of grass etc), Inclination of the track (Track gradients in different reaches of the track), number of sharp turns and undulations, etc. These aspects can be taken into consideration by fixing an appropriate Track Index for each track, as well as for each distance on that track. This Track Index could be defined as the distance in meters travelled by a horse rated 150 on the TF scale carrying 120 lbs weight (54.5 kgs) in a one minute period on that track, assuming the track to be at its fastest on the day. It may be noted that a particular racetrack could have different Track Indices for different distances particularly for tracks having different inclinations in different reaches. For this purpose, all the track characteristics such as the geometric layout indicating the profile in plan with radii of all curves, the undulations and gradients, and the ground characteristics (nature of soil, specially the top layers), etc. etc. are required to be known and understood. However, its not worthwhile gong into such scientific research and analysis, such as understanding the ground characteristics, ground inclinations, etc and trying to evaluate their impact on reducing the speed of horses. Instead of that, it is better to perform a statistical study of the past few years of the standard timings for various distances for various classes of races, in particular the record timings in order to arrive at the long-term track index.

I have done this research for the Bangalore racetrack, and observed that the timings for shorter distances (1100m) are relatively unimpressive by about 2.5 to 3% when compared to those of middle distances and those of longer distances (2400m) relatively unimpressive by about 1 to 1.5 % when compared to middle distances. What does this mean: Obviously there is a certain downward gradient from some point near the middle-distance zone for about half the track perimeter and a corresponding rise in the remaining half. To some extent the sharp curves may also contribute to this. These assumptions were more or less found to be accurate when after extensive search on Google I came across the following para pertaining to the Bangalore racetrack:

"The Bangalore Racetrack is an oval-shaped, right-handed course measuring approximately 1950 meters with four sharp curves and pronounced gradients. The downhill backstretch drops 13.10 (43 feet) meters from 1800 meters to 800 meters and climbs 11.53 meters (38 feet) from that point to the winning post with a further rise of 1.5 meters (5 feet) from the winning post to the 1800 meters. This demanding and testing race track, with its gradients, bends and a distinct short straight, places a premium both on the speed and endurance of the horse and the skill and experience of the jockey; a win on this race track is therefore a significant achievement"

When I correlated this info with my earlier findings and assumptions I was able to work out the Long Term Track Index (let's call it `Ti`) for various distances for the Bangalore race track, which I give below:

1100M : 1048m, 1200m :1055m,1400m : 1066m, 1600m : 1077m, 1800m :1082m, 2000m: 1077m, 2400m: 1072m, 2800m : 1066.

Please do not ask how the above values have been worked out from the given gradients of the course. There are enormous complexities involved. The methodology used is purely empirical (and not based on a scientific study of motion along inclined surfaces). In broad terms I have considered a gain in time @0.45secs per 100m from 1800m to 800m, a loss in time @ 0.55 secs per 100m from 800m to the winning post, and @ 0.3secs per 100m from winning post to 1800m. In addition a loss in time of 0.75 secs for each semi circle curve is also considered. It has also to do to a considerable extent with the correlation between standard timings and ratings for various grades/classes of horses, it goes without saying that some intuition was also necessary in arriving at the final figures.

2) DUE TO SHORT TERM TRACK INDEX

The Short term track index is a measure of the Penetrometer reading values that determine the day to day variations in the track conditions caused by rain and storms.

For the Bangalore race track the short term track index (say T_s)values may be arrived at by reducing from 1090 in the following way

a) For PR reading (say `p`) less than 2.5cm: No reduction required.

b) For `p` between 2.5 and 3.5 reduce Track Index by (p – 2.5)*5

c) For `p` between 3.5 and 4.0 reduce Track Index by 5 + (p – 3.5)*10

d) For `p` between 4.0 and 4.5 reduce T_s by 10 + (p – 4)*25

e) For `p` between 4.5 and 5.0 reduce T_s by 22.5 + (p – 4.5)*35

f) For `p` exceeding 5.0 reduce T_s by 40 + (p – 5.0)*50

As a general rule for all race tracks the adjustment in timings for short term track index can be taken as: 1.0 for Perfect going (Fast track), 0.998 for Good going, 0.995 for Good to Soft ground, 0.99 for soft going, 0.98 for Heavy going, 0.97 for very heavy going, and 0.96 for yielding ground.

Having fixed the criterial for adjustments in timings due to both categories of track indices for the Bangalore Racetrack, it is possible to calculate the Absolute value of the Time form rating of a horse based on a specific performance at the Bangalore racetrack. This is best explained by a couple of examples of some notable performances of the recently concluded Bangalore summer season.

1) NORTHERN LIGHT ..Race no: 57, $W = 59.5$kg, $D = 1600$m, Timing $= 95.25$ seconds. PNR $= 3.0$ Adjustment factor for T_s (Short Term Track Index) $= 0.998$, Adjustment factor for T_i (Long Term Track Index) for 1600m $= 0.98807$, Adjusted Timing $T = 95.25*0.998*0.98807 = 93.93$, Timing T_1 over 1600m for a 150 rating $= 90.7$s.Hence Time form Rating $= 2.22*W + 30 - 22222*(T - T_1)/D = 2.22*59.5 + 30 - 22222* (93.93-90.7/1600 = 117$

2)   ALL ATTRACTIVE ..Race no: 102, $W = 60$, $D = 1200$, Timing $= 71.57$s, adjustment factors 0.998, and 0.96789, adjusted timing $= 69.13$, $T_1 = 66.4$ and TFR $= 2.22*60 + 30 - 22222*(69.13 - 66.4)/1200 = 113$

UNCERTAINTY ELEMENT:

It should be noted that the rating does not take into account:

a) The reserve potential of the horse, ie how handy it was near the winning post or

b) The pace of the race – the fractional timings etc - ie whether it was a slow run race or a fast run, or

c) The fitness of the horse.

Factors `a`, `b`, `c`, referred above, including many other factors correspond to the uncertainty element of the race. It is here that the caliber or the expertise of an observer can make a difference while giving a figure by which the timing of the horse must be adjusted, to take into account all the uncertainties.

The most important aspect to be remembered is that even if you are the greatest of experts in the line and have made an absolutely perfect evaluation of the Time form rating of a horse even to the nearest 0.5 point, there is no guarantee that the next time around the horse will run true to its rating. At best you can only fix probabilities of success.

As in most cases the uncertainty element is generally quite high, it is advisable to determine the absolute value of TFR for only a select few horses only where this element of uncertainty is minimal in the eyes of the observer. For all other horses the TFR values may be determined by comparative studies only and for this purpose closely fought races only must be considered.

SOME CASE STUDIES:

In a similar way I worked out the `Ti` value for the Epsom Derby course for the mile and a half as 1056, which gives fairly accurate results.

Based on 1056 as the LongTerm Track Index for Epsom over the Derby Distance, the Time Form Rating of WORKFORCE (fastest timer) as per his Derby Win some years back is worked out as:

TF RATING OF WORKFORCE:

Distance travelled `D`: 1mile 4f + 10 yards =2424m

Time taken: 2min 31.33 secs.

Track Index `F` for Epsom Derby course: 1056m

Adjusted time `T` = 151.33*1056/1090 = 146. 61 secs.

Time T1 (corresponding to a 150 Points TFR for distance 2424m = 142.2*2424/2400 = 143.83 secs

Weight carried `W` = 56.75 kgs.

Hence Time form rating = 2.22*56.75 + 30 − 22222(146.61 -143.83)/2424 = 130.50 say 130

Simiarly I worked out the `Ti` value for the Ascot Race Course for the mile distance as 1040, which also gives fairly accurate results. Based on this Long term Track Index I have worked out the Time form Rating of the June 2021 St. James Palace Stakes winner ``POETIC FLARE`` as given below:

TF RATING OF POETIC FLARE:

Distance travelled `D`: 7Furlongs 213 yards = 1603m

Time taken = 1min 37.4 secs.

Long term Track Index for the course and distance: 1040m

Adjusted time: `T` = 97.4*1040/1090 = 92.93 secs

The going on the day (track condition) was good to firm hence the short term track index was 1090, hence no adjustment in timing required on this account.

Time `T1` corresponding to 150 points TFR for distance 1603m is worked out as 90.7*1603/1600 = 90.87secs

Weight carried `W` = 9.0 st = 56.75 kgs

Hence Time form rating = 2.22*56.75 + 30 – 22222(92.93 - 90.87)/1603 = 127.43 say 127

As I mentioned earlier the Time form Rating does not take into account the reserve potential of the horse, ie how handy it was near the winning post or

the pace of the race – the fractional timings etc - ie whether it was a slow run race or a fast run. This aspect is reflected in the fact that on the same day as Poetic Flare won the St James Palace Stakes at Royal Ascot, Palace Pier who is actually higher rated than Poetic Flare won the Queen Anne Stakes= in 99.18 secs over 8F thus getting a rating of only 110 as worked out below:

Hence, adjusted time = 99.18*1040/1090 = 94.63sec, `T1` corresponding to a TFR of 150 = 90.7*1610/1600 = 91.27 secs. W = 56.75kgs

Time Form Rating = 2.22*56.75 + 30 – 22222(94.63 – 91.27)/1610 = 109.6 say 110

Conclusion: While Poetic Flare ran to almost full potential, Palace Pier might have had something in hand.

Regarding Time Form Rating of FRANKEL, as I mentioned `Frankel` was rated at 147 by the British Experts, I am not very sure it was correctly evaluated. He never clocked timings which would justify such high rating. They say his two best performances were his win at Royal Ascot in the Queen Ann Stakes in 2012, and his win in The 2000 guineas in 2011. In the former He clocked 97.85 seconds over 8 f (1610m). There are at least six horses who ran faster than that in the QUEEN Ann viz Valixir 96.64s in 2005, Ramonti 97.21s in 2007, GoldiKova 97.74s in 2010, Toronado 97.73 in 2014, Ribchester 96.60 in 2017, and Lord Glitter 97.40 in 2019. Even though Frankel won by more than 8 lengths and must have had something in hand, there is no calculation to show the extent of this reserve potential. If we watch the race more closely, it can be seen that even though he had a big lead, he was still being ridden fast and did not slow down too much. Perhaps he could have taken one second less, At best equalled Ribchester`s time of 96.60 seconds. Considering the Long Term Track Index adjustment of 0.955 for Royal Ascot over 1600m, the adjusted timing $= 0.955*96.6 = 92.25$. The timing over 1610m corresponding to a 150 TFR $= 90.7*1610/1600 = 91.27$.

Weight carried was 56.75Kg, accordingly the TFR $= 2.22*56.75 + 30 - 22222*(92.25 - 91.27)/1610 = 142.5$ which in my opinion is also on the higher side.

Frankel`s 2000 guineas win was already rated at 140 by the Time Form Organisation.

It is still not clear what was the calculation for the 147 TFR given by the British Experts

While, still on the subject and still continuing with examples of case studies, I would like to say that I once spent a lot of time trying to analyze the 2006 performance of 'Deep Impact` in a race called `Teno Sho` at the Kyoto race course (Japan) where he ran 3200m in 3m 13.4sec, an unbelievably fast time for that distance. I was trying to evaluate his Time Form rating based on that showing. The Japanese public rates this performance as better than Secretariat's showing in the Belmont, but the UK experts did not rate him higher than 134 which is curious. In a conversation with some guys on YOU tube in the posting `Deep Impact – Tribute`, I had commented that such a time was impossible to believe unless a portion of the track which is covered twice during the 3200m run is having a net downward inclination of substantial quantum in that portion. and in my quest to determine the TF rating of Deep Impact, someone from the JRA (Japanese racing authority) actually helped me in this connection by giving valuable inputs (Refer `Deep Impact – Tribute` on YouTube). To my surprise, someone from the JRA itself responded promptly to my comment. His response given below:

`@SKSAGAR100 From the JRA - "features a flat home straight of just over 400 meters. The course also is known for its hill laid out over the third and fourth turns, where the track rises from the 1,200-meter mark to the 800 mark, then dips right to the mouth of the stretch. The trip for the Tenno Sho lasts one-and-a-half laps around the oval." The outer oval is 1894m. Go to the JRA site, it has charts with course details (overview, undulation, etc.). How does it effect your figures?`

I accessed the site and found the layout and the geometrics. In some portion of the track which is covered twice in the 3200m run there is a rise of 1 in 90 for 360m followed by an extremely sharp fall of 1 in 40 for 160m.This last part could rejuvenate a tiring horse to a good extent. Further interaction with a certain `711ATOM`presumably from JRA on `YouTube` revealed there are three more occasions when a sub 3m 15sec time has been registered over 3200m at the Kyoto Race track. Compare this with the World Record time of 3m 16.4secs clocked by `Kingston Rule`in the Melbourne cup, and the time of 3m 18secs over 2 miles (3218M) by `Petrone` at Hollywood Park in 1969. With all these inputs and again with a great deal of `intuition` - as these inputs are far from sufficient (I have asked for the course record for 1600m over the outer and inner oval tracks which info, if it comes will be quite useful) - I am inclined to give a long term track index of 1122(about 3 % higher than a perfectly fast track) to the Kyoto Race track over 3200m.

Accordingly the adjusted time of Deep Impact is worked out as 193.4*1122/1090 = 199.07 secs and the Time form rating of Deep impact is worked out as:

$$58*2.22 + 30 - 22222(199.07 - 196.7)/3200 = 142.3$$

This puts Deep Impact in the Elite group of race horses rated 140 or higher. Though he may not figure in the all time top ten list, he still has to be the highest rated horse of the 21st century so far, at least 3 points better than Sea the Stars or Harbinger. But the UK Experts could not rate him higher than 134 which is curious. No doubt he lost in the French Arc, finishing 3rd. But I have no doubt that on his home turf in Japan he would have beaten the best in the world.

It appears that the UK Time form rating experts might be a bit biased in favor of European horses. Even some top-class American horses are rated lower than their actual worth. What I am surprised at is: Why is there no scientific system in place - acceptable to all countries and turf authorities - that takes care of all the technical parameters and arrives at the accurate values of the ratings. It's true there are complexities, but it's not Quantum Physics. All that is required is an accurate evaluation of the Long term track index (defined earlier) of the track for various distances. This can easily be done either by experimental observations over a period of time or by a detailed statistical study and analysis of the past performances on the track.

Having dealt at length on the methodology of evaluating Time Form Rating of a horse based on a specific Performance, the question may be asked, how to make use of the system to predict winners, here are some ideas which might be of some use.

It is best If I explain with examples:

CONSIDER THE SANTA ANITA RACE TRACK (DIRT)

Let us proceed in steps :

STEP 1: Obtain the Track Record timings for various distances - easily available and in public domain - and determine the Time Form rating for each performance assuming the Standard Long Term Track Index as 1090, as explained below:

a)   Distance(D) : 5 f (1006m), Horse : Wildfire Kid, Draw:6 (add 1m To D), Weight (W):115 lbs (52.3 kg). Timing (T) :56.34 seconds. T for 1000m = 56.34*1000/1007 = 55.95, TF rating = 2.22W + 30 – 22222 (T – T1)/D = 2.22*52.3 +30 -22222*(55.95 -54.7)/1000 =118.4

b)   D: 5.5f (1107m), H : `I Am The Danger`, Draw:5, W:117(53.1kg),T= 61.64 s, T for 1100m = 61.19s, TFR = 2.22*53.1 +30 -22222*(61.19 -60.45)/1100 = 131.5

c)   D: 6F (1208M) `The Factor` Dr:2 W:118(53.5 Kg) T =66.98, T for 1200 = 66.54s, TFR = 2.22*53.5 + 30 - 22222*(66.54-66.4)/1200 =145.8

d)   D: 7F (1409m), `Twirling Candy` Dr:3 W:118 (53.5kg) T = 79.70, T for 1400m = 79.14s, TFR = 2.22*53.5 +30 - 22222*(79.19-78.4)/1400 =135.5

e)   D:8F (1610m), `Ruhlmann`Dr:4,W:118, T=93.4s, T for 1600 =92.82s, TFR = 2,22*53.5 + 30 -22222*(92.82 - 90.7)/1600 = 118.7

f) D:9F (1811m), `Star Spangled`, W: 117 (53.1 kg). T = 105.8s, T for 1800m = 105.2s, TFR = 2.22*53.1 + 30 - 22222(105.2-103.2)/1800 =122.3

g) D: 10F (2012m), `SPECTACULAR BID`, W: 126 (56.75 kg), T =117.8s, T for 2000m = 117.8*2000/2012 = 117.1s, TF Rating = 2.22*56.75 + 30 – 22222(117.1 – 115.9)/2000 =142.7

h)

i) D: 11F(2213M), `Be Faithful` W: 55.5kg, T= 135.20s, T for 2200m = 134.4s, TFR = 2.22*55.5 + 30 - 22222*(134.4 -128.8)/2200 = 96

j) D: 12F(2415m), `Queen`s Hustler`, W: 51kg, T=147.2s, T for 2400m = 146.28s, TFR = 51*2.22 + 30 - 22222*(146.28 -142)/2400 =102.4

k) D: 14F(2818m), `Noor`, W: 52.7Kg, T = 172.8s, T for 2800m = 171.7s, TFR = 2.22*52.7 +30- 22222*(171.7 - 169)/2800 = 125.6

STEP 2 : Determine the Long Term Track Index for the various distances

A close look at the above TFR values for various distances reveals a complex picture. Its excellent for 1200m, very good for 1400m, reasonably good for 1100m and 2000m, not so good for 1800m and 2800m, pretty bad for 2400m and even worse for 2200m. How do I make sense of this? What kind of layout of the track can account for such variations as seen above?

That is the question.

Now, Let us have a look at the Track Layout which is in the public domain and is reproduced below:

Course map

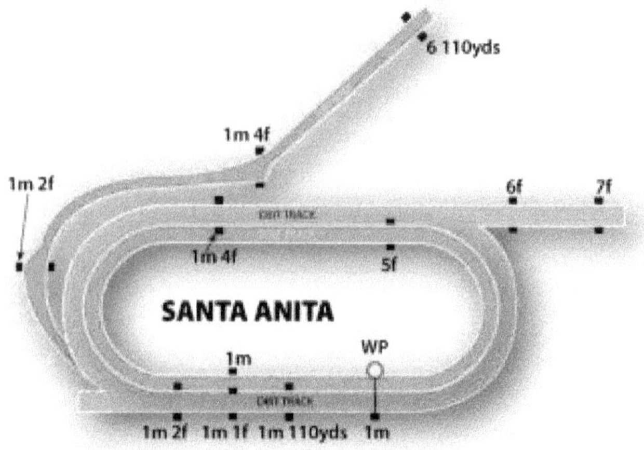

The oval dirt track is one mile round with a run-in of 330 yards.

There is no information on the track gradients, inclinations etc.

Now if I - with a small team - could walk on the complete track with a levelling instrument at our disposal, I would be able to prepare a comprehensive plan of the geometric layout with all rises and falls throughout the track length.

In the absence of that, some calculated guesswork and some intuition will be required.

Let us begin with the 5f say 1000m race. It looks as if there is a level surface for about 400m (ie 600m from the winning post), and this is followed by one complete semicircle of about 300m, and in this stretch there is also an appreciable rise in level, then there is a sustained fall till the winning post. I am inclined to give a Long-term Track Index (Ti) value of 1080 for this distance (1000m). This will give Wildfire Kid`s performance a Time Form Rating of about 131. It looks as if there is a sharp fall in level from 1200m to 1000m which is responsible for the excellent timing of the `The Factor`. I suppose we can give a Ti value of 1100 for this distance, which makes `The Factor`s performance a TFR of 134. For 5.5f race the Ti value can remain as the standard 1090, hence `I am the Danger` retains his TFR at 131.5.

From 7f to 6f there might be a slight rise in level, Ti for this distance can be about 1095, giving a TFR of 130 to `Twirling Candy`

For the 1 mile race (1600m) which is the same as winning Post after traversing the full perimeter of the oval shaped track, hence there can be no level difference. However, there are two sharp semi circles to be maneuvered. A Ti value of 1080 is recommended, giving `Ruhlmann` a TFR of 130.5.

It is likely there is a fall of some appreciable value from the 2000m (same as the 400m) point to the 1600m (same as the wp) point. I recommend Ti values of 1090 and 1085 for the 2000m and 1800m distances respectively thus giving TFR values of 142.7 and 128.5 to Spectacular Bid and `Star Spangled` respectively.

Coming to the 2200m and the 2400m distances, in both cases there are three semi circles to be maneuvered, in two of which an appreciable rise is encountered. I recommend a Ti value of 1070 for these, thus giving a TFR of 122 to `Be faithful` and 128 to `Queen`s Hustler`.

Finally in the case of the 2800m distance, there is an advantage of one additional fall from the 2800m point to the 2600m point (which is also the 1000m point), a Ti value of 1085 is recommended, thus giving `Noor` a TFR of 132.

Hence The Long-Term Track Indices for various distances at the Santa Anita Dirt track are - as per my understanding - as given below:

1000m: 1080, 1100m: 1090, 1200: 1100, 1400:1095, 1600m: 1080, 1800m: 1085, 2000m: 1090, 2200m: 1070, 2400m: 1070, 2800m: 1085.

STEP 3

Get hold of the race book of the relevant Racetrack/Centre and determine the Time Form Rating of as many horses as you like depending of course on the time at your disposal. Perhaps it will be good idea to do it for about 50 to 60 horses of the two highest classes apart from about 20 decent 3yr olds.

It is advisable to start with a few select fast run races, with close finishes where horses ran to their full potential. In this way you can determine the appropriate TFR rating for about 20 to 30 horses, for the remaining horses the TFR values can be worked out by comparative studies.

SOME GUIDELINES :

Racing has several variables which can neither be predicted nor quantified. I mention a few here:-

(a) A horse's physical and mental being on that day.

(b) A jockey's physical and mental state on that day.

(c) Effect of the going

(d) Effect of a new equipment.

(e) Effect of a new distance. Most horses have not raced over 2400 m. before the Derby. Is the rating arrived at a mile valid over a mile and a half ?

(f) Getting or not getting a level start.

(g) Getting or not getting a good position.

(h) Pace of the race.

(j) Interference during the race, not getting an opening at the right time.

(k) Horse losing a shoe.

(l) Jockey losing the whip.

(m) Jockey misjudging the pace, not timing his challenge precisely

As mentioned earlier, in most cases the uncertainty element is generally quite high, hence it is advisable to determine the absolute value of TFR for only a select few horses where this element of uncertainty is minimal in the eyes of the observer. For all other horses the TFR values may be determined by comparative studies only and for this purpose closely fought races only must be considered.

Again, as mentioned earlier, the most important aspect to be remembered is that even if you are the greatest of experts in the line and have made an absolutely perfect evaluation of the Time form rating of a horse even to the nearest 0.5 point, there is no guarantee that the next time around the horse will run true to its rating. We cannot expect horses, trained by humans and ridden by humans, to conform to our mathematical calculations. We can only expect them to perform within a "zone". Accordingly this aspect of a `Mathematical Approach` is just one of the factors to be considered but not the deciding one.

At best we can only fix probabilities of success, whereby we can select a few horses (2 or 3) in a field and fix their probable chance of success (Win or Place or Show etc) by giving appropriate weightage to the various aspects including of course the `Mathematical approach` apart from some intuitive thinking (hunch), and then look at the odds available, play the percentage game and determine which of these is a `Value bet` and play accordingly..

....

SLIP FORM

It might be a good idea to play – on you tube - and listen to a song by Hank Snow called `The Last Ride` while reading this short piece. It's a `Talking`, `Singing`, `Talking` `Singing` etc kind of music, and you can imagine yours truly doing the `Talking`, Singing`, `Talking` the article on the same tune as `The Last Ride`

TALKING:

Now this one is from the late seventies.

We were in Bangalore …our office on the road called Residency.

A call from the Corporate office Kolkata .. it appeared there was an urgency

The caller .. an Executive Director asking a Structural Engineer to be sent immediately.

To a Thermal Power Plant in a place called Singrauli …

to resolve a technical problem and to solve it very quickly …

And the chosen structural engineer was none other than yours truly.

Now my entitlement was first class by train, but due to urgency

Air fare was sanctioned .. but only for the onward journey.

So I flew from Bangalore to New Delhi… and from New Delhi to Varanasi

And at Varanasi the company jeep was waiting for yours sincerely

To take me from Varanasi … to Singraulli.

And all the time I had no idea what the problem was technically

So I was in some sort of stress ..both physically

As well as mentally

In short the stress was both tensile as well as torsional

The only consolation was that a very senior level engineer (Not the overall boss) at this Project Singrauli

Was known to me personally … his name US Tiwary

And it was by name that he had requisitioned for me.

SINGING:

*The jeep went straight to the project site and not the guest house booked for me.

My friend Mr. US Tiwary was waiting there for me.

We shook hands and wasted no time and started going up the PIERS

He went up the ladder in no time at all while I was still downstairs.

The vertigo and the tiredness was having its toll on me

But I guess I had to start climbing as there was no choice for me.
*

TALKING:

So I slowly started climbing that narrow cat ladder … what else could I do

I took my time .. and reached the top at nine .. to the `how do you do`

With Piers on either side I saw a massive Slip form

And there was congestion all around …

as slow speed concreting was going on

With all kinds of construction materials lying all around

SINGING:

*I asked the obvious question: `What exactly is going on wrong ?`

There is too much congestion I was told and chaos all around

The problem .. concreting going so slow .. so a solution had to be found

To speed up the work .. before I was Bangalore bound.

So I lingered on the platform for some time while looking around

And tried to understand how a slip form worked. *

TALKING:

My friend Tiwary who was a specialist in slip form construction explained how a vertical slip form relies on the requirement of a quick-setting concrete, but at the same time it needed to be workable enough to be placed to the formwork, gained early strength and became sufficiently strong so that the form could slip upwards without any disturbance to the freshly placed concrete.

There are hydraulic jacks that support the system and lift the forms at a predetermined rate to match the speed of construction. These Jacks had to be strategically located on the working platform with steel rods attached to the worker`s scaffolding, and these steel rods would get embedded in concrete.

The site visit was followed by a dinner meeting where we were joined by a Swedish Guy (I don`t remember the name) who was from a Swedish company that supplied the slip form system to HSCL (the name of our company). I could sense that the two guys .. UST and the Swede .. were not exactly on friendly terms .. But I was able to generate a technical discussion .. Some brainstorming ensued .. I learnt a lot during the discussions.

And then I came up with an idea … How about a Two Tier Platform .. With an upper tier about 1.25m wide on each side of the main platform of the system and at a height of about 1.85m (if I remember correctly). All the reinforcement rods .. which occupied a substantial space .. would be placed on the upper tiers .. and the reinforcement guys could conveniently install and fix them at the appropriate locations in the piers as per the drawings. This would relieve the congestion on the main platform in a big way, and likely speed up the concreting works to some extent.

The Idea was accepted by the two of them.

The slipform system would have to be redesigned. I said I`ll provide the conceptual scheme before Lunch on the next day, and then return to Bangalore, arrange to prepare the detail fabrication drawings and send them to Singrauli at the earliest.

I took a set of drawings of the existing Slipform system from the Swedish Guy. Mr.Tiwary dropped me at the guest house in his jeep.

I looked at the drawings and spent about half an hour thinking about the conceptual arrangement I would be working on next morning before I went off to sleep at 1.30 am.

Next morning after BF, I sat in UST`s site office, and worked on the structural design of the supporting system for the upper Tiers of the revised slipform system, understood the revised loading pattern and redesigned the Jacks .. the numbers required and their most appropriate locations .. The rods to be embedded in concrete .. and the overall framework.

UST joined me .. after his morning site visit .. he went through my design .. suggested a few changes here and there and together we prepared freehand sketches (two copies) of the GA of the entire set up.

I signed the sketches and told UST that the fabrication drawings for the revised system will be issued from Bangalore within 24 hrs of my return back. But he looked at the sketches and said this is enough for me, and that I can send the drawings at leisure.

In the afternoon we met our Chief Engineer along with a very senior engineer from the client`s (NTPC) side and discussed the proposal. They were pleased. The client insisted I should stay on at Singrauli and prepare the fabrication drawings in all respects before returning to Bangalore, but my friend UST said that won`t be necessary, that he will manage to fabricate and erect the system based on the sketches.

I left Singrauli the same evening .. spent the night at a Hotel in Varanasi .. could not see much of the town .. as I experienced one of my life`s worst headaches.. perhaps due to exhaustion.

I spent the ensuing weekend (two nights) enroute at New Delhi.

When I returned to Bangalore on the forthcoming Wednesday, I was informed that the new slip-form system was already erected and operational, and the average concrete per day was already up by 60 % on day 1 and as much as 80 % from the 2nd day onwards.

Our Executive Director from Kolkata .. during his visit to Bangalore .. called me to his chambers at UTILITY BLDG. to congratulate me for a job well done.

For a few days I was sort of a hot topic of the company

Our Chief Engineer at The Singrauli Project Office sent a letter of appreciation for me.

I still have it with me

And its like a trophy.

Thank you

… …

ABOUT ART AND MY ARTIST WIFE

ART APPRECIATION ... AT AN ART CAMP

The other day, some years back, I was invited as a chief guest at an art camp for young artists and asked to say a few words on Art Appreciation. My qualification for being the Chief guest: Husband of a prominent artist (Bharati Sagar) from Bangalore and author of the deeply philosophical book `SIX WORDS` which had nothing much to do with art.

After the usual formalities ... and some fumbling ... I came to the main topic and said something to this effect:

``Regarding Art Appreciation, now I am not an artist by profession, nor have I done a course in art or art appreciation. So, what do I say?... It is true one does not have to be a batsman to realize that Don Bradman or Tendulkar were great batsmen ...or...one does not have to be an actor to realize that Laurence Olivier was a great Actor. But I guess it is not the same thing with art. Art is way too complex, in particular, the measurement of the quality and value of art. Indeed one has to be very much an artist to realize that Da Vinci was a great artist and Mona Lisa is a great work of art. For, an artist who has spent a number of years studying and practising art, knows and understands that Art too has some rules and techniques that need to be followed, and that there is Science in art too, and it is complex. A look at the university Syllabus of a major art course will give an idea about the complexities involved.

In my view the true or the intrinsic value of art is a measure of the extent to which the experience affects us in an emotional sense.

How people value culture is subjective and involves making judgements about quality based on how it makes them feel. This can include our feelings of connection to the artwork and our own personal subjective opinions of its quality based on the way it makes us feel. … Which means ..apart from pleasing to the eyes .. it must also be pleasing to the mind. It should provide the connection to the mind and to the feelings. At the same time, I do not like to analyse the painting too closely and intently and thus lose sight of its beauty and its appeal to my senses.

And even for an artist, this requirement of connectivity holds good, I think, when making his own measurement on the quality of art. It is not so simple to distinguish between great art and very great art. Beyond a certain limit. say 9.5 marks out of 10. It is not so easy to pinpoint a precise value to a work of art and say that such and such work is valued at 9.7 or 9.6 out of 10. It becomes subjective. This is of course my view, and I could be wrong.

But to drive home the point I am trying to make, consider a thought experiment. Assuming nine different portraits of men and women carried out by some of the leading artists of the world, are displayed in a gallery alongside Mona Lisa, assuming further that the great Van Gogh comes to the gallery to judge all the ten portraits, and that he has neither heard the name of Da Vinci nor about Mona Lisa.. what are the **chanc**es that he will give the highest rating to Mona Lisa ? I guess about 10 to 15 %. If all are equally good technically, then perhaps he will look for a connection, and maybe select the portrait which is matching with his style of painting.

I hope I am not misunderstood. I do not wish to underestimate the value of great works of art. They are indeed great, and we should admire and be deeply inspired by the great artists who created them, but not necessarily `overawed` by them. I reiterate, that in this crazy world, it is distinctly possible that a great work of art `A` which may be just about as good as or perhaps slightly (1 %) better than another great work of art `B`, but somehow has a market value several thousand times more.

And I wish to say, that, each one of us present here (except yours truly) has the potential to become a great artist.

Coming to the Art Camp, this is indeed something unique and wonderful, and provides a platform to the young artists and art lovers to meet and interact with the senior artists of Bangalore, and in the process learn a few things about art and art appreciation and to make good use of that knowledge by incorporating the same in their daily lives as well as to understand on what aspects they should focus on so as to obtain better satisfaction and enjoyment from what they do.

As for my advise to the young artists looking for a bright future .... in Six Words I`ll say : `` Be inspired and be deeply connected ``

The inspiration need not be from a great Artist alone ... it could be from a great Musician, a great sportsman, a great Scientist, a great religious or spiritual leader, even a great politician (less chance), a great novelist, ..or even a great theory.

The connection ... I believe ... has to be much more wide ranging .. connection to the one who inspires, connection to your thought process, to the ideas that come to the mind, the ideas you wish to convey to the world, to the rules and techniques of art, and above all to the subject of your choice... and the deep extent to which you should go into the understanding of that subject.

The connection ... that provides you with a driving force.. that gives you the energy .. the passion to work .. the exhilaration as your work proceeds and so on.

What can be that subject ?... well there are several thousand interesting subjects to choose from ... choose the one that interests you the most.. as far as possible it should be an uncommon subject ..so that there is a good chance that it provides you with a unique identity linked to that subject.. Examples of some uncommon but very interesting subjects:

``The oneness of the mind``.. ``Interconnectedness of the universe`` ``The self-destructing human civilization`` ``The nuclear arsenal available on the planet Earth`` ... ``The current geopolitical situation on the planet`` , ``Disruptive technology`` , ``Biomedical warfare`` , `Convergence of Science and religion`` , ``Eastern Mysticism`` , ``Mysticism as Art, etc etc to name just a few.

Having decided on the subject, go deep into it, read books, study and analyse, run your imagination and develop your own ideas. I tell you, its good fun. And It is up to you to what extent you want to incorporate your thoughts in your painting, it can be realistic or it can be abstract, its your choice (It is only artists who have this freedom of choice .. Scientists don`t have it), make it mysterious looking, difficult to understand by the public, or an art critique. The less they understand, the more they will be curious and the more they will keep looking at it.

Or you can select a common subject. about which you might have some uncommon but very interesting ideas of your own

I can go on and on, but I guess I should stop

Thank you.

ARTIST`S STATEMENT .. By BHARATI SAGAR .. PRIOR TO AN ART SHOW

I have liberated my art from the parameters of the known, to the paradigms of the intangible and ephemeral spaces of consciousness that define and express a world, where the individual is surpassed by the whole.

It takes contradictions and paradoxes to create forms in a continuum of energy that embraces the construction and destruction of forms of life, as in the dance of Shiva.

Influenced by the thoughts and writings of my husband Surendra Kumar Sagar, my work prior to this was a narration of human endeavour in the microcosm that is the means of touching the macro levels of the time and space reality that we as human beings on this planet experience.

It is the committed work of Surendra to bring awareness to the mind, individually and collectively, to the responsibility that we have towards this larger life of the planet Earth and its inhabitants. In the present context he traces the story of nuclear destruction and proliferation and the dangers of devastation in the future.

I take his ideology to the level of art, spontaneously flowing from my mind to brush & canvas as these works of art. I hope to lift the viewer from the gravitational force that conditions and restricts our perceptions, to a new space where we become united in a common heritage and vision that is timeless and eternal.

For the painting `**Travelling Cosmic Mind**` and my new series, I have experienced with unconventional materials like lemon juice, industrial waste metal dust etc. I have hence - as mentioned earlier liberated my art from the parameters of the conventional methods to create new forms that embrace the construction and destruction of forms of life as in the dance of Shiva.

My paintings influenced by the thoughts and writings of my husband Surendra Kumar Sagar, trace the story of the evolution of the `Intelligent Field` in the cosmos in the form of an omnipresent `Cosmic mind`. We all have the same mind, except that our consciousnesses are each in the singular. The Cosmic mind is within us and empowers us to observe and participate in the affairs of the universe, which in turn are based on the interactions of the world on us. The Intelligence in the field is however a measure of the `Information` in the field. Sometimes the Intelligent Field gets contaminated with evil information, such as conflicts and wars, nuclear disturbance and proliferation and the dangers of devastation we face in the future.

We together endeavour to bring awareness to the responsibility that we have towards the larger life of the planet Earth and its inhabitants. I thus take this ideology to the level of art, flowing from my mind to brush and canvas, and I hope to lift the viewer to a new space where we become united in a common heritage and vision that is timeless and eternal.

BHARATI SAGAR

ALWAYS PROUD OF HER

While she was at work on these paintings, I could sense the passion and enthusiasm in her approach, I could also see the extent to which the experience affected her in an emotional sense, in terms of her feelings of connection to the artwork as well as to the ethos of the book Intelligent Field. How people value art is subjective and involves making judgements about quality based on how it makes them feel. In my case I could feel that connection .. which means .. apart from pleasing to my eyes .. the paintings were also pleasing to my mind. At the same time ..I do not like to analyse the paintings too closely and intently and thus lose sight of their beauty and appeal to my senses.

I am proud to say that the exhibition – just like all her exhibitions – was well appreciated. It was the talk of the town particularly in art circles. Indeed, it was a success story.

MY UNDERSTANDING OF HOW ART CAN BRING ABOUT CONVERGENCE BETWEEN SCIENCE AND RELIGION

Just as Science and religion must converge to end conflicts and wars, so must Art and Science converge in order to bring about that convergence between science and religion.

Before I proceed further, let me explain about my understanding of religion, and then I shall endeavour to explain how Art can be utilized to bring about this convergence.

So, what is `Religion`?, I do not mean `What is this or that religion?`, I mean `What is the essence of religion`? In my view, the essence of religion is four fold:

a) To find solace and peace of mind in prayer and meditation.

b) To live and let live.

c) To be understanding and compassionate.

d) To take care of the environment around us.

In essence there is nothing `Hindu`, Muslim` or `Christian` about Religion, just as there is nothing `American` or `Indian` about Democracy, or nothing `Greek` about the Pythagoras Theorum.

Next what is Philosophy ?

What can that philosophical model be?

I believe there are three initial conditions that need to be satisfied in order to develop the concept of a philosophical model which is acceptable to all.

The established principles of science must not be violated.
The universe must make sense.

The question regarding existence of God and the form in which God exists must have just one answer. It cannot be, that a theologian has a different answer from a naturalist, or a Scientist. The truth is only one. What can the truth be which is acceptable to all and which also satisfies the two conditions mentioned above?

Having said that I will now try to contemplate what I believe can be a model of philosophy that is acceptable to most if not all and at the same time satisfies the three initial conditions listed above.

The central theme of my philosophical model is `Unity of Consciousness` as explained in the six words : ``We all have the same mind``. It's just that our consciousness is in the singular, but it's the same mind everywhere… like an Infinite Cosmic mind that gives us eternal consciousness. We can call it an `Intelligent Field` that is forever on the lookout for a certain biochemistry to give itself consciousness and then understand itself as well as understand the universe, and also understand that the universe makes sense. Whether we call it `Cosmic Mind` or `Intelligent Field` or `God`, is just a matter of taste. .

But, how do we arrive at the convergence and how do we do it through Art.

But, how do we define Art? As for me I define art as nothing other than `contemplation`.. as defined by the French Artist `Augeste Rodin`, a contemplation as the joy of an emotional comprehension of the universe and the ability to recreate by illuminating it with consciousness. And as the great Mystic Ruysbrueck said ``*Contemplation places us in a purity and radiance which is far above our understanding*``

In short Scientists try to understand natural phenomena, artists contemplate and explore reality and create work for literary and aesthetic enrichment, and the general public tries to utilize the output of both art and science for their functional and spiritual needs.

So we contemplate something, based on the Interactions of the world on us and the information stored in our brains consequent to those interactions, and develop a picture of it in our mind and draw and paint it on the canvas, or compose a symphony, whatever. To the artist and to the audience it must be pleasing to the eyes as well as pleasing to the mind, and perhaps make sense. In addition, it should strike a chord, Sometime the striking of the chord can create visions that can inspire scientific discoveries. For example, Niels Bohr attended an art exhibit where he saw a cubist painting in which Picasso showed his mistress' face in both profile and face front. Bohr wondered which was the real portrait. Then, AHA! both views were the real portrait, depending on the viewer's point of view. This led to Bohr's Complementarity Theory and the beginning of quantum physics.

The main reason why I think `Art` can be instrumental in bringing about a convergence is that Artists can reach out to the common man better and faster than Scientists. A Scientist has to go through a whole lot of reviews and rejections before even a semblance of a theory, or even philosophy gets accepted. To appreciate Science, one requires some understanding of it. We must know Geometry if we wish to understand trigonometry. But we need not learn the language of music, to be moved by a Symphony, or the technique of Art to be moved by a painting.

During the 2013 WOMEN`S ECONOMIC FORUM, my wife Bharati Sagar gave a Talk on `Convergence of Science and Religion Through Art` Prior to visiting New Delhi for her talk, she interacted with, some of the leading artists of India and apprised them of the subject of her talk and requested them to send their views on the subject and if possible send images of their work which corroborate to the idea of this convergence. The response she received was very very encouraging. She received lots of these images of their art, with interesting captions.

A video was prepared and displayed on the big screen while she was giving the talk. Later, it was uploaded on You tube. Here is the link for the same:

``Convergence of Religion and Science Through Art ... Curated by Bharati Sagar``

....

GENERAL EXPRESSION FOR DETERMINATION OF WARING LIMITS

(Written in November 2014)

`` In this article a general expression has been conjectured which gives the number `x` being the maximum number of n th powers required to represent, when added together, any finite integer. In other words, there is no positive finite integer which cannot be represented as a sum of `x` nth powers. ``.

Study and research done during summer of 1959 while studying for Post graduate course in Mathemetics in Bombay University.

Article first written in 1960 ...not published

Article updated in 1986, and sent with a covering letter to `The Director...Institute of Mathematical Sciences... Madras on 4th April 1986... It was not published but an encouraging response received.

Article recomposed in November 2014 and tells of my letter to IMS Madras, the manuscript of the 1986 article, and compares it with the latest inputs available on `Google` on the subject. The November 2014 article is being published in the website.. `www.sixwords.in`

In April 1986, I wrote a letter to The Director, Indian Institute of Mathematical Sciences, Madras on the subject `Edward Waring`s conjectures (1770)` on theory of numbers, inviting his kind attention to an article in Illustrated Weekly of India, relating to the recent proof of Waring limit for 4th powers by Dr. R. Balasubramanium in collaboration with two French Mathematicians. It appeared – with that proof – that Waring limits up to 9th power stood proved, at the time. It was also mentioned - in the weekly – that there is no documented evidence available which shows how Edward Waring arrived at his conjectures 200 years back.

In my letter I mentioned that I had derived a general expression for a number `x` in terms of `n`, such that there is no positive integer existing which cannot be represented as a sum of - not more than - `x` n^{th} powers of integers. The derivation of the general expression was entirely based on known Waring Limits up to the 7th power of which I was aware of at the time. However – I wrote – that the authenticity of the general expression can be a gauged from the table – which I enclosed – which gives the value of `x` as worked out from the general expression for up to `n` $= 15$, by tallying the same with corresponding Waring limits up to 15th power which should be available with the latest literature on the subject. The importance of the general expression can be judged from the fact that if this is proved, it will act as a comprehensive proof of Waring`s Conjectures for all powers.

The referred table which constituted a part of the letter gave the values `x` for all values of `n` from 2 to 15, as 4, 9, 19, 37, 73, 143, 279, 558, 1079, 2132, 4223, 8384, 16673, and 33203 respectively.

MANUSCRIPT OF THE ARTICLE OF APRIL 1986 :

``Waring limits postulated over 200 years back (in 1770) by Edward Waring have once again been in the limelight following the recent proof of the Waring limit corresponding to the 4th powers by Dr. R. Balasubramanium of the Institute of Mathematical Sciences in collaboration with French Mathematicians Jean Yuves Peshouillers and Francois Press.

What are these Waring limits?

Waring`s theorems or rather conjectures, since he could not prove them, are on the following lines :

a) Every integer is either a square or it can be represented as a sum of two, three, or upto four squares. In other words no positive finite integer exists which cannot be represented as a sum of four or less than four squares. The number four can be considered as a Waring limit for squares.

b) Every integer is either a cube or it can be represented as a sum of two, three, four, or upto nine cubes. The number nine can be considered as a Waring limit for cubes.

c) Every integer is either a fourth power, or it can be represented as a sum of two, three , four..or upto 19 fourth powers. The number 19 can be considered as a Waring limit for 4th powers.

d) Likewise Waring conjectured that every integer is either a 5th power, a 6th power, a 7th power etc, or it can be represented as a sum of upto 37 5th powers, 73 6th powers, 143 7th powers etc respectively, and the numbers 37, 73, and 143 being regarded as Waring limits for 5th, 6th, and 7th powers respectively.

As can be seen, there is no distinct pattern existing between the Waring limits for various powers (4, 9, 19, 37, 73, 143, etc), and to date there is no documentary evidence available which shows how Waring arrived at his postulations. During the last 50 years or so, Mathematicians have proved Waring`s Theorems upto several powers, the latest one being that of Dr. R. Balasubramanian's proof corresponding to 4th power. However, it appears there is no record of a successful research having been carried out for the evaluation of a generalised expression for Waring limits, let alone its proof. The author here has conjectured to derive a general expression based on available Waring limits upto 7th power. Using this expression waring limits upto 15th power are determined and indicated in table `A`.

EVALUATION OF GENERAL EXPRESSION FOR DETERMINING WARING LIMITS :

A detailed analytical study of the Waring limits upto 7th power will reveal the following:

a) The smallest integer that requires the summation of the maximum number of squares is the number 7, and it requires one square of 2 and 3(*) squares of 1.

b) The smallest integer that requires the summation of the maximum number of cubes is the number 23, and it requires 2 cubes of 2 and 7(*) cubes of 1.

c) The smallest integer that requires the summation of the maximum number of 4th powers is the number 79 and it requires four 4th powers of 2 and 15(*) 4th powers of 1.

Proceeding in the same way, it can be found that the smallest integer requiring the summation of the maximum number of 5th powers is the number 223 (requiring six 5th powers of 2 and 31(*) 5th powers of 1), for 6th powers the corresponding number is 703(requiring 10 6th powers of 2 and 63(*) 6th powers one), and for the 7th powers the corresponding number is 2175 (requiring 16 7th powers of 2 and 127 (*) 7th powers of 1.

The numbers marked with an asterisk contained within brackets (*) ie 3, 7, 15, 31, 63, and 127, are significant, and follow a distinctive pattern, each being one short of 2, one short of cube of 2, one short of 4th power of 2, one short of 5th power of 2, one short of 6th power of 2, and one short of 7th power of 2 respectively.

In each of the above cases it is observed that the smallest integer that requires the maximum number of nth powers is less than the nth power of 3. Denoting this smallest integer as `D`, it is seen that `D` requires the summation of nth powers of 2 and 1 only.

Hence D = a*2^n + b*1^n .... (A1)

Where `a` and `b` are constants.

Further it is noticed that `D` is in each case found to be exactly one short of the number containing the maximum number of nth powers of 2 before the nth power of 3.

Hence by definition: D = c*2^n − 1 .... (A2)

Where `c` is a constant giving the maximum number of nth powers of contained in 3^n. ie c is equal to the positive integral value (quotient only) of the expression (3^n/2^n).

For n = 2, c = quotient of the expression (3^2/2^2) = 2, and D = 2*2^2 -1 = 7

For n = 3, c = quotient of the expression (3^3/2^3) = 3, and D = 3*2^3 -1 = 23

For n = 4, c = quotient of the expression (3^4/2^4) = 5, and D = 5*2^4 -1 = 79

For n = 5, c = quotient of the expression (3^5/2^5) = 7, and D = 7*2^5 -1 = 223

For n = 6, c = quotient of the expression (3^6/2^6) = 11, and D = 11*2^6 -1 = 703

For n = 7, c = quotient of the expression (3^7/2^7) = 17, and D = 17*2^7 -1 = 2175

As indicated earlier `D` contains n^{th} powers of 2 and 1 only. Denoting by `x` the maximum number of nth powers required to represent D, it is obvious that x is equal to the sum of the quotient (ie the positive integral value) of the expression $(D/2\wedge n)$ and its remainder `R`.

Ie x = quotient of $(D/2\wedge n) + R$

From equation A2 $D = c*2\wedge n - 1$

Hence x = quotient of $|(c*2\wedge n - 1)/2\wedge n)| + R$

Substituting for c as quotient of $|3\wedge n/2\wedge n|$

Where $|\ |$ denotes the positive integral value (quotient only) of the expression within.

We get x = $\{(2\wedge n*|3\wedge n/2\wedge n| -1)/2\wedge n\} + R$ ....
(A3)

Which is the general expression for Waring limits.

Where { } denotes the positive integral value (quotient only)of the expression within, and R as the remainder.

Thus it can be stated that every integer is either an nth power or it can be represented as the sum of 2, 3, 4....or at the most `x` n^{th} powers, or in other words there exists no positive integer which cannot be represented as the sum of `x` n^{th} powers, where `x` is given by the expression in (A3).

Based on the expression in (A3) the values of `x` as determined upto 15th powers ie for n= 2, 3, 4 ...etc ...upto n =15 are 4, 9, 19, 37,73, 143, 279, 548,1079, 2132, 4223, 8384, 16673, and 33203 respectively.

A more simplified expression for `x` than that indicated in (A3) can also be determined.

As per (A1) : D = a*2^n + b*1^n, and x = a + b

As found out earlier, the value of `b` is just 1 short of 2^n, ie b = 2^n - 1 and the smallest integer D contains 1 n^{th} power of 2 short of the maximum number of nth powers of 2 that can be contained in 3^n, ie a = | 3^n/2^n | -1

Hence x = a + b = | 3^n/2^n | + 2^n - 2 .... (A4)

Where, | | denotes the quotient (ie the positive integral value) of the expression within.

Expressions (A3) and (A4) above, give identical results. This can be checked and verified for all values of n from 2 to 15.

CONCLUSION :

On the basis of known Waring limits upto 7th power, a general expression has been derived for `x` in terms of `n`, such that there is no positive integer that exists, which cannot be represented as a sum of x n^{th} powers. The primary assumption which facilitated the derivation being that the smallest integer `D` that requires the summation of maximum number of nth powers is less than 3^n, and that D is in each case just one short of the number containing the maximum number of nth powers of 2 before 3^n. In the validity of these assumptions lies the proof of the general expression which in turn would prove Waring`s conjectures comprehensively.

.... Manuscript ends ....

UPDATE ON THE SUBJECT :

Following my brief interaction with Institute of Mathemetical Sciences , Madras (now called Chennai) in connection with the manuscript, I tried hard with my search to find the latest literature on the subject in particular on the general expression, but couldn`t find anything worthwhile, there was no Google at the time. And then, I got busy with work –Design and construction of structures – and later got myself deeply involved in Science and philosophy, and the article was out of mind ever since for the last 28 years. In the interim I along with my team of Structural Engineers received an International award ..`The Bentleys Global Be Inspired Award` for innovations in Structural Engineering for the year 2013, and wrote a book on Science and Philosophy called ``SIX WORDS... Seminar held in a parallel universe``, published in January 2014.

Now – in November 2014 - I am including the manuscript in my website, to solicit reader`s comments and thought I should find out the latest on the subject ..Through `Google` of course.

During my search, I found an article on `Recent Progress On Waring`s Theorums And Its Generalisations ` ...BY L. E. Dickson.

Its a very lengthy write up... here are two small extracts from the article :

`` ***On the Ideal limit for Universal Waring Theorum*** : It has been proved that every positive integer is a sum of 9 cubes, whence 9 is the ideal limit for cubes. For biquadrates (fourth powers) the ideal limit is 19.``

`` *The ideal limit for* $n^{th}$ *powers is* $I = q + 2\char`\^n - 2$, *where q is the greatest integer less than* $(3\char`\^n/2\char`\^n)$ ``

Dickson goes on to with his elaborate study and analysis to discuss various theorems related to the subject pertaining to the Ideal limit for various powers. His analysis is extremely complex and difficult to comprehend. It is also a little bit confusing and has a few errors too; In his table of Ideal limits he gives a value of 558 for the Ideal limit for 9th powers, which according to me is 548 (perhaps it might be a misprint).

I could not find in Dickson`s analysis or anywhere else the simplified methodology to arrive at the general expression put forward by me in my article.

I have prepared an excel sheet for evaluation of `x` (the ideal limit in Dickson`s nomenclature ) for various values of `n`. Proceeding further the values of `x` for n values 16, 17, 18, 19, and 20 are determined as 66190, 132055, 263619, 526502, and 1051899 respectively.

....

BRIEF OUTLINE OF BELL`S INEQUALITY:

Two sets of measurements need to be carried out on a pair of quantum level particles that were in contact sometime, but are now separated. The entangled quantum state is written down for the four spin $\frac{1}{2}$ particles and the expectation value of the products of certain binary measurements performed on the individual particles is calculated. It is then worked out on the basis of the probability theory, that if local causality is considered while determining the expectation value, it produces a contradiction.

This contradiction is called Bell`s Inequality. If this contradiction is satisfied, it confirms locality. If this contradiction or inequality is violated, it confirms non local action.

In the book `Cosmic Code`....by Heinz Pagel.... Two experiments have been described, one for classical level objects – Special nail guns shooting nails sideways with their long axis perpendicular to the direction of motion – where Bell`s inequality is satisfied confirming local causality, and one for quantum level particles – Positronium atoms decaying into photons and then travelling in opposite directions – where Bell`s inequality is violated confirming non local action.

Recall the great adventures of Sherlock Holmes, or the books of Agatha Christie, Edgar Wallace, or Dan Brown or any other book which caused you to say `Wow-that was clever`, and then judge for yourself if that quantum of cleverness could match the quantum attributed to Bell and his inequality, and can we not attribute this adjective called `Clever` in equal measure to Roger Penrose and

Heinz Pagels in capturing these ideas and putting them across in such an ingenious manner, and also in equal measure to Andrew Whittaker in first understanding and then describing a complete historical review of the past sequence of ideas from Broglies `Pilot wave ` concept to Bohm`s `Hidden variable` concept etc and many others leading upto Bell`s ideas.

Many experiments were actually carried out by the Scientists in the Eighties, and in 1982 an experiment carried out by the French Scientist `Alain Aspect` on photons up to a separation distance of about 15 km successfully demonstrated that Bell`s Inequality gets violated at quantum levels, thus confirming `Non local action` or `Quantum entanglement`.

Quantum theory stood vindicated. However the questions whether Quantum theory is complete or incomplete still remains unanswered. But the EPR Paper which was designed to prove the incompleteness of quantum theory as it led to a paradox called `Non local action` could not prove it , instead this `Non-local action` or `Quantum Entanglement` is here to stay and emerges as another feature (Mind boggling no doubt) of Quantum Mechanics. Indeed we have to live with this `Action at a distance` whether spooky or not spooky.

But can we make use of this QE to send messages at speeds faster than light?

Coming back to the adjective `Clever`, let me tell you that none is cleverer than the God who plays dice. He has played a marvelous trick on us by avoiding real time non local influences. If you read the referred Cosmic code chapter carefully and also a number of times, then at some stage and at some point of time, a certain flash of understanding may strike you and you will realize that the answer to the above question is:

NO … We cannot.

The argument that is used for this purpose is that:

A random sequence altered randomly remains a random sequence

At this stage I will not go into the details except to conclude that:

God is a Mathematician.

Indeed `Mathematics is the President and the name of the company is `The Universe`

Mathematics designs the universe by creating laws and tells Physics to execute them…

Laws such as the `Uncertainty Principle` where `Mathematics` - the President - has played such a trick that it does not allow the quanta to get created out of nothing except for the shortest possible and thus irrelevant period of time.

Laws such as `Quantum entanglement` where `Mathematics` – the President – has played such a trick that it permits entanglement but does not permit information to be sent faster than at speed of light.

And both these tricks have been played out by the Mathematician by incorporating a certain randomness in the nature of reality… In the former case ..the vacuum randomly fluctuates between being and nothingness…and in the latter case the Mathematician keeps shuffling the deck of nature in such a way that the randomness remains intact . I should go as far as to say that `Randomness` is the Vice President.

And then at the center of everything ... there is this thing called `EQUIVALENCE` ..The cleverest of all the laws ...it goes without saying that this law is out and out Mathematics and Physics is just dancing to its tune.

Who won the debate, was it Einstein or was it Bohr?

This question too remains unanswered still. At present all we can say is :

Neither Einstein nor Bohr lost the debate

To the question whether QE has any practical applications, `mindboggling` is the word that comes to the mind. Scientists of the category called `Indifferent` are already working on it and Quantum computation is looming large on the horizon. It could also hasten the collapse of civilization.

....

DARK REALITY AS I SEE IT

"We all live day to day completely oblivious to the fact that we're a part of a much larger and stranger reality than we can possibly imagine."

— Blake Crouch, <u>Dark Matter</u>

``Perhaps nothing confounds our understanding of galaxies so much as the realization that the gas and stars that we observe constitute only a small fraction of the total mass of all large structures of the universe, from galaxies on up.`` --- Mitchell Begelman and Martin Rees, Gravity`s Fatal Attraction

The Universe as I see it is expanding and I expect it will keep expanding as long as the density of matter in the universe is less than a certain critical density. Scientists have worked out the value of such critical density as about 5 atoms per cubic meter of space. They have also worked out that the current density of actual matter (Baryonic matter comprising of Protons, Neutrons and Electrons) in the universe is only about 0.2 atoms per cubic meter of space. But if this is true then the universe should be expanding at a much faster rate than the current rate at which it is expanding. And the scientists have also worked out that the current rate at which the universe is expanding corresponds to a matter density of about 4 atoms per cubic meter of space. So, what is it that makes up the difference in matter density between 0.2 atoms per cubic meter and 4.0 atoms per cubic meter of space?

We do not know. Let us call it `DARK MATTER`.

Who discovered Dark Matter?

In the early 1930s, while studying Hubble's observations of the Coma Cluster of galaxies, a scientist named Fritz Zwicky noted an anomaly. According to the measure of visible mass, single galaxies were moving too fast for the cluster to remain bound together. Zwicky realized that the aggregate mass of all the galaxies we could see in the Coma Cluster was inadequate, by a wide margin to prevent the system from flying apart.

Further evidence for Dark Matter lies in the fact that our galaxy Milky Way and the Andromeda galaxy, which are the dominant members of a Local group of galaxies are falling toward each other despite the fact that the total mass of the baryonic matter in the group is insufficient to prevent the impending collision of the two galaxies which could happen in 4 to 5 billion years.

WHAT IS DARK MATTER MADE UP OF?

I think it might be Neutrinos.

Unlike the photons, which are massless, it is possible that neutrinos have a non-zero mass. If they have the right value for their mass based on the total number of neutrinos (and antineutrinos) that exist, they could conceivably account for 100% of the dark matter. Now let me do some calculation on what the right value should be - or near about - of the mass of neutrinos in order for them to qualify as 100 % contributors to dark matter. To do this I need to corelate the subject with the rate of expansion of the universe. For the universe to keep expanding at the current rate the amount of matter available in the universe should not exceed about 4 atoms per m ^{3} of space in the universe. But we know that the amount of normal matter is only about 0.2 atoms per m ^{3} of space in the universe. So, for the neutrinos to contribute the remaining 3.8 atoms per cubic. meter, it transpires that the total mass of all the neutrinos is about nineteen times the combined total mass of all the protons, neutrons, and electrons in the universe. Now how many protons are there in the universe? I take this figure as about $3.0*10^{80}$ (As suggested by Arthur Eddington). The mass of each proton is about $1.672*10^{-27}$ kg .So the total mass of all the protons in the universe is about 5×10^{53} kgs. The number of Electrons is the same as number of protons, the mass of each electron is about $9.11*10^{-31}$ kg, hence the total mass of electrons is worked out as about $2.8*10^{50}$ Kgs . The number of Neutrons is about $0.5*10^{80}$ and their total mass $= 0.5*10^{80} \times 1.675 \times 10^{-27} = 0.8*10^{53}$ Accordingly the total mass of available matter in the universe (excluding dark matter) is about $3.8*10^{53}$ Kgs. Which means the total mass of dark matter (neutrinos) is about $19*3.8*10^{53}$ kgs ie about $7.3*10^{54}$ kgs. Mass of each neutrino is neglible but not zero. I take it is about $0.6*10^{-36}$kg

That makes the number of neutrinos - should they be considered as exclusive members of the dark matter club - as about $4.8*10^{90}$

WHERE DID THE NEUTRINOS COME FROM?

They came from Supernovas. With just two minutes left in the life of the star, there was a stupendous struggle against gravitational collapse and neutrinos were created in phenomenally large numbers, and they escaped through the star into space leaking energy into space thus not permitting energy to support the star against collapse.

Leaking Energy Into Space …

This is interesting to me ...

Let us corelate this with some other aspect of this `Dark Realty` .. Let us call it `Dark Energy`

DARK ENERGY

``There had to be some unknown and mysterious energy out there in the cosmos pushing everything apart, fighting against Gravity. What could this powerful energy be? Why had it never been seen before? What generates it, and where was it hiding? The theorists' extraordinary conclusion was that this unknown energy came from the very vacuum of space``. (Universe & Cosmology & Laws & Gravity & Space & Dark Energy) Horizon: From Here to Infinity, BBC 1999

If we think of galaxies as masses, we can postulate: To exist, they are removing energy from the vacuums that surround them. We know from Einstein's famous mass energy equivalence ($E = mc^2$), that removing energy is equivalent to removing mass. This suggests that as the mass/energy density decreases, the gravitational attraction within the vacuum decreases, which in turn causes the vacuum to expand.

This still leaves two questions:

The concept of `vacuum energy` acting as `dark energy` that drives the expansion of the universe is interesting. Roger Penrose has discussed this subject in great elaboration in his latest book `Cycles of Time

Lets talk `inflation` ...the cosmological one.. Was it the `vacuum energy` that caused a rapid increase in the size of a certain scale factor ?

` `*The inflation of the scale factor meant that a small, smooth spatial region of the universe expanded exponentially to encompasses a volume that would grow to become larger* `TODAY` *then the size of the observable universe. In the process of expansion, the spatial geometry became flat* ` ` .. `New Physics`, edited by Gorden Fraser.

Consider the word `TODAY` in Gorden Fraser`s Para (actually the para is by W.L. Freedman and E.W.Kolb) ... For us on the Planet Earth `TODAY` corresponds to about 13.7 billion years since the big bang and we say with our highly advanced, and at the same time highly limited understanding of the subject, that sometime during the first second (between 10^{-36} secs and 10^{-32} secs to be precise), after time zero of the big bang, the universe inflated to a size (Radius) of $10^{`z`}$ metres, where `z` is about 28. Now imagine someone living on a certain planet orbiting a certain star in a certain galaxy which is about 12.3 billion light years away. Considering that I am looking at him now, it's obvious that his universe is only about 1.4 billion years old. For that someone, if he or she is writing a book on Cosmology, `TODAY` corresponds to only 1.4 billion years since the big bang, and in his description of `Inflation` `z` will not be 28 ..rather it will be closer to 27. In the same way, for someone ahead of us in the deep future z` will be greater than 28.

What then was the true extent of the inflation with regard to the volume encompassed?....Was Einstein`s special relativity actually violated by the `entities` that travelled with the `inflation`?...

Or can we say that `special relativity – in respect of giving an upper limit to speed of light - ` is not violated, in the same way as in Quantum entanglement when we say that entanglement takes place instantaneously but no information can be sent faster than at speed of light ?

Or is this value of `z` a measure of the consciousness of an observer as well as a measure of his coordinates in time and space at the time of his observation ?

Or is the concept of `Inflation` itself somewhat misconceived?, and should we now look at some other explanation for the homogeneity of the universe ?

Or can we say that `There is indeed a physical law that `prohibits` space expanding faster than the speed of light`?

No one really knows the answer. It could be that space expanded faster than the speed of light. This judgment is consistent with the accelerating universe we observe today.

For all practical purposes the big bang with all the activities associated with it – including inflation – is still in a state of superposition. Its wave function (I mean the `Probability wave` function) has not collapsed as yet. Will it be collapsed any time in the future, I should think so, If it really was such a stupendous show - as it is made out to be by all and sundry - it could not remain a performance to empty stalls. The designers must have made sure that in course of time it will be witnessed by the consciousness of the observers of the future.

But it has not been observed so far.

Some scientists are of the view that the size of the observable universe is about 93 billion light years in diameter, what exactly is implied? Does it mean we can observe regions which are say 20 billion light years away, which means we can look back 20 billion years into the past and enroute somewhere we can pierce thru the big bang itself. Surely the capacity of telescopes could not have been a constraint, we have observed Supernovas that happened billions of years back in much detail, why not a few percentage point extra capacity and the big bang should have been observed clearly considering the enormity of its magnitude. But the truth is.. it is not observed so far.

What then is the explanation?

Perhaps:

a) The size of the observable universe is limited to the age of the universe in terms of light years which works out as about 10^{28} meters radius.

b) Beyond this we cannot see no matter how great may be the capacity of our telescopes, as that would mean going back into time beyond `Time zero` into a different universe altogether or may be to a previous aeon of our own universe, or maybe the main show of the big bang took place in a black hole from which light could not escape at all.

Or maybe the big bang was caused by some sort of a phase transition in the last stages of the previous aeon of our universe. This required a huge violation of the second law of thermodynamics in bringing about a comprehensive change in entropy from an infinitely high level to a near zero level. Perhaps the Super consciousness of the previous aeon played a part in this transformation.

Perhaps the massless photons of the last stages of the previous aeon were not getting bored at all, they were always upto something, getting together and forming a `Super consciousness'. Designing the new universe with brand new and refined constants and then switching on the big bang.

Now, the big bang and inflation cosmology is very well understood and accepted by the Scientific community, but it has not been physically observed – through a telescope that can look at a region 13.7 billion light years away - nor adequately explained so far. As for inflation, it cannot be tested as the energies involved are too high and well beyond the experimental reach of our accelerators. When I say `The big bang` has not been adequately explained, it is implied that an appropriate ` PHYSICAL CAUSE` has not been determined (which is by and large acceptable to the scientific community) so far, of which the `EFFECT` is precisely what happened at all the various stages of the big bang, not just the first second of it.

And that `physical cause` Even if it happened just Plank time before the Previous Aeon ended, must not violate any law of science, and it is further assumed that all the laws of Science which are applicable in the present universe were also applicable in the previous Aeon.

Now what can be that possible `Physical cause` that explains it?

What can be that `Hidden variable`?

Regarding Inflation, this can be dispensed with, by assuming that the previous Aeon ended with a size that permits Homogeneity, and by considering that the big bang was a phase transition that converted `something` into `temperature, radiation, energy, etc, whatever it was that came with the big bang.

All that remains to be answered is, what was that `something`?

There is however not yet a proper explanation of why the vacuum should possess an energy with the complex property of `negative pressure`, even though in quantum theory there is a precedence for vacuum energy arising from virtual particle-antiparticle pairs.

But there is a large scale mis-match between the amount of energy thus acquired and that which is just sufficient to blow the cosmos apart. In principle this is precisely the mismatch between Quantum Theory and the relativity based cosmic space time model, which needs to be reconciled, and for that we have to wait till a perfect understanding of Quantum Gravity is achieved, till such time we may have to consider the topic of Dark Energy as a work in progress.

There is no doubt however that cosmic expansion of the universe is caused by the Dark Vacuum energy and this is predominantly the main contributor (perhaps 75 %) to the overall energy-density of the universe. It does not however contribute towards matter-density for which I believe the main contributors are Baryonic matter (protons, neutrons, electrons) and Dark matter (Neutrinos). The former contributes about 5% and the latter 95%.

So, it looks like there is Dark Energy that is driving the expansion of the universe and there is Matter (mostly Dark) that is restricting the expansion to a reasonable limit which is just about right (As per the Designer`s requirement) to : 1) Prevent the tearing away and bursting apart of the universe and 2) Prevent the collapse of the universe on itself in a big crunch. In other words .. just about right for Life and consciousness to exist and flourish in the universe.

BACK ONCE MORE TO THE NEUTRINOS

Some scientists believe that the speed at which the neutrinos travel could be a wee bit faster than the speed of light. There was a blog `CERN: LIGHT SPEED MAY HAVE BEEN EXCEEDED BY SUBATOMIC PARTICLES` By the

Scientists Frank Jordan and Seth Bornstein, on Huffington Post dated 22nd September 2011.

My comments on the blog: There are many reasons that may have led to the expectation that neutrinos may travel a wee bit faster than light, such as:

Even in a vacuum, creation and annihilation of particles/anti particles goes on all the time which may cause some obstructions to the photons of light, but none to the neutrinos. Speed of light may not be a true constant of nature, but changes value as the universe evolves, but there may not be any change in the speed of neutrinos. Light neutrinos of mass – energy less than 1 Mev may be non-relativistic now, ever since they decoupled themselves from interacting with other particles. As the lightness increases, so does the ability to travel faster. Or as some respondents have pointed out, neutrinos may go through another dimension.

…… ……

ANSWERS TO QUESTIONS OF QUIZ TIME

A1: DEMATERIALIZING

A2: 1225

(38 X 27 = 1026 for the word `Dematerializing` + 14 x 3 = 42 for the word `Devolution` + 9 x 3 =27 for the word `Islander` + 12 for the word `Languish` + 23 for the word `Zone` + 15 x 3 = 45 for the word `Grumbling` + 50 points for using all seven letters = 1225)

A3: WE ALL HAVE THE SAME MIND

A4: 4 & 1

Einstein`s number was 5 which is possible in two ways, ie 3 + 2 = 5, and 4 + 1 = 5. In the first case, the corresponding Newton`s number would be 6, which could be either 6 x1 = 6 or 3 x 2 = 6, which means he is not sure which are the two numbers. In the second case, the corresponding Newton`s number would be 4, which could be either 4 x1 = 4 or 2 x 2 = 4, which means in the second case also Newton would not be sure which are the two numbers. Einstein knows this and that is why he could tell Newton: `I ***know that you do not know the two numbers***`.

Newton`s number was 4 which is possible in two ways, ie 2 x 2 = 4 , and 4 x 1 = 4. In the first case, the corresponding Einstein`s number would be 4, which could be either 3 + 1 = 4, or 2 x 2 = 4. In the former case the corresponding Newtons number would be 3 in which case Newton would be sure that the two numbers are 3 and 1. Thus if Einstein`s number was 4, he would know that it has a possible combination of 3 and 1 where Newton would know the two numbers, in which case Einstein could not have made the statement: `I know that you do not know the two numbers`. But when Einstein did make that statement, Newton at once knew that Einstein`s number was 5 and not 4. Hence Newton knew that the two numbers (corresponding to Einstein`s 5`` and his own `4`) could only be 4 and 1, and hence Newton made the statement : **`In that case I know which are the two numbers`**

Meanwhile, Einstein had already worked out that Newton could make his statement only if Newton`s number was 4. For, if Newton`s number was 6, it could also correspond to 6 x 1 = 6 which makes Einstein`s number as 7. So if Einstein`s number was 7, he could still have made the statement `I know that you do not know the two numbers`, but Newton would not have found that statement or hint useful to him, and he would have kept quiet. But Newton did not keep quiet, he made the statement `In that case I know which are the two numbers`. Einstein Knew that Newton could make that statement only if his number was 4, which corresponding to Einstein`s number 5 was possible only if the two numbers are 4 and 1. Accordingly Einstein could make the statement ``**In that case even I know which are the two numbers``**.

A5: 325

A6: 36 %

A7: I DO NOT KNOW HOW TO DECODE IT

A8: 135500

N= number of years elapsed between 1979 and 2030 = 51

Y = CAGR = 20*0.99462^N = 20*0.99462^51 = 15.190

V = 100*(1 + Y/100)^N = 100*(1.15190)^51 = 135543

A9: You and the Pope are two different people. But $2 = 1$, hence you are the pope.

A10: from `TWO TO TWO TO TWO TWO

A11: CHICAGO

A12: 18

A13: INFATUATION

A14: 75 kms per hour

A15: Jane Kaha Mera Jigar Gaya Ji, Abhi abhi yahin tha kidhar gaya ji, Kisi ki adayon pe mar gaya ji, badi badi akhiyon se dar gaya ji

A16: UPSET

… … …

SCIENCE RELIGION AND THE INTELLIGENT FIELD

(Presentation at the International and Interdisciplinary Conference at Kolkata)

Imagine a ``travelling cosmic mind`` that can ensure a permanent everlasting consciousness for us.

This short piece – Not too short - aims to draw the modern-day scientist`s and philosophers' attention to the idea that there is an `Intelligent Field` out there that controls nature but the amount of intelligence in the Intelligent Field is a measure of the `Information` in the field. The deeply philosophical discussion is aimed at merging Eastern and Western philosophies together into a cohesive philosophy and arrive at a `Standard Model of Philosophy` that is likely to find acceptance as a foundational principle, to resolve the conflicts of the world.

Now, a Standard Model, such as that of `Particle Physics` or `Cosmology`, in general signifies a model which has been developed, as far as possible, on the basis of the established principles of Science.

In which case, you might ask the question: How can the said term `Standard Model` be used for philosophy?, for philosophy is after all philosophy, if it were an established truth, it will not be called philosophy.

In fact what exactly is the meaning of the word `Philosophy` ?

Let us use some high school algebra to understand this.

If `x` corresponds to the total ultimate truth and `y` corresponds to the portion of truth that is fully understood and established by Science, then `z` = x – y obviously corresponds to that portion of truth which is either unknown or partly known and not yet established by Science. The entire philosophy of the world – whether written in books/websites, or discussed in seminars/get-togethers – or in `Talks` like this, is dealing with `z`, and that portion of it which is related to such topics as `mind`, `consciousness`, `soul` , `spirit`, and `God` is dealing with `Religious Philosophy`. As and when a portion of `z` becomes an established truth by Science (By experiment/observation), it gets added to `y`, till such time, it remains in the domain of philosophy/religion.

In my opinion the singular criterion to be adopted in arriving at one`s religion is that it should be fully compatible with `y` and should not be required to reject an already established truth.

People`s beliefs are dependent on the interactions of the world on them, and these beliefs change with time, again based on the IOTW. The truth however is only one. Its not easy to know what is that truth. The best that we can do is to think of something which has the greatest probability of finding acceptance by all religions, and the greatest probability to lead towards convergence of science with religion. Merely saying that `x` percentage are Atheists, `y` percentage are Protestants, `z` percentage are Catholics…and some are Hindus, Muslims, Buddhists doesn`t serve any purpose. There must be `Something` out there that controls nature, that we can look up to, that assures everlasting awareness/consciousness to all of us. The standard model of philosophy must aim at reaching an understanding of what can be that `something` out there which controls nature, and what can be that `understanding` which is acceptable to most, if not all sections of society.

So, what can be that understanding?

And how can we understand consciousness?

UNDERSTANDING CONSCIOUSNESS AND TELEOLOGY

I believe understanding consciousness is still a work in progress, in particular that category of `understanding consciousness` which is linked with teleology. Regarding the other category of `understanding consciousness` which considers consciousness as a discrete phenomenon from neural activities, there are various groups of scientists who have tried to explain the subject in different ways and a lot of important and credible work has been done on the subject.

But I am not a member of that club.

I like to discuss and understand the subject from a teleological perspective even at the cost of appearing `Out of place` here. In my view even the best theories about the nature of the universe are so far unable to reconcile the very small and the very big. Under these circumstances we cannot rule out the possibility that the only way to round up our understanding may even require `TELEOLOGY` or something similar like `ENTELECHY`.

But then I still need to be scientific about it. It cannot be that there is Magic in Physics. It cannot be that there is an `Act of God` involved.

So, I try to explain the whole thing in a different way, as I have done in my books ``SIX WORDS`` and `INTELLIGENT FIELD`

I summarize my ideas as follows:

1)   The essential material reality is an Intelligent Field.

2)   The field obeys the principle of special relativity and quantum theory.

3)   The field links the quantum with the classical.

4)   The field looks for a biochemistry to give itself consciousness to understand itself.

5)   The intensity of the field at any point is a measure of the Information in the field.

6)   There isn`t anything else.

Now, I expand and explain all the five lines one by one :

LINE 1 : The Essential Material Reality Is An Intelligent Field

I take it that the `INTELLIGENT FIELD` is the ultimate boss of the universe made up of fields, such as the `HIGGS FIELD` that provides mass, the PROTO-CONSCIOUSNESS FIELD` that provides consciousness, the ELECTROMAGNETIC FIELD that provides light, the GRAVITATIONAL FIELD that provides influence and the `INFORMATION FIELD` that provides information about temperature, about density, etc etc, and in general about what's going on.

LINE 2: The Field Obeys the Principle Of Special Relativity And Quantum Theory

I take it that this is the Initial condition that must be fulfilled, that all the laws of science to the extent that they are established principles of science are obeyed and are not permitted to be violated.

LINE 3: The Field Links the Quantum With The Classical

Before I proceed further, I provide some quotes by Scientists which have relevance here:

``Decoherence effects are the basis both of the mechanism whereby our thoughts can affect our actions, and of the reconciliation of quantum theory with our basic intuitions`` ... ``The quantum state of the brain is reduced to a collection of `parallel potentialities, each of which is `essentially` a classically conceivable possible state of the brain`` ... ``your physically described brain is an evolving cloud of essentially classically conceivable potentialities``...Henry Stapp (from the book `Information And The Nature Of Reality .. Edited by Paul Davies and Gregerson

``Probabilities do not arise at the minute quantum level of particles, atoms, or molecules – these evolve deterministically – but seemingly, via some mysterious larger-scale action connected with the emergence of a classical world that we can consciously perceive``....Roger Penrose (from `Emperor`s new mind`)

To explain the above:``Consider that I am walking along a road and need to turn left at the next turning to go to my destination....Now the quantum entities inside my body keep moving here and there `at random`, with complete disregard to the phenomena of cause and effect, but they will in their trillions be at the right places at the right times to ensure that I am turning left and not going straight or turning right. And in case something happens to me before I turn left, such as I get hit by a car or something, the classical world (of the car hitting me) informs the quantum world of the changes in probabilities, and the quantum particles inside me will be at the right places at the right times to make sure that I fall down, provided of course, that the impact of the car was sufficient for the purpose.``

I take it that the Intelligent Field acts like a guiding force that links probabilities from the classical levels to the quantum levels. Its a hugely mysterious connection

LINE 4: The Field looks for a Biochemistry to give itself consciousness to understand itself.

In quantum mechanics particles do not have a specific shape or location, until they are observed or measured. Is this a form of proto consciousness at play? According to the late scientist and philosopher, John Archibald Wheeler, it might be. In his view, every piece of matter contains a bit of consciousness, which it absorbs from this proto-consciousness field. He called his theory the "participatory anthropic principle," which posits that a human observer is key to the process. Perhaps the universe may be conscious, and the intelligent field might be like a Proto-Consciousness Field. Roger Penrose supports the concept of Panpsychism, the idea that the universe may be conscious.

According to Penrose: `The laws of physics produce complex systems and these complex systems lead to consciousness, which then produces mathematics, which can then encode in a succinct

*and inspiring way the very underlying laws of physics that gave rise
to it.* `

Penrose goes on to say: `*I do not believe that we have yet found
the true* ` `*Road To Reality* `, *despite the extraordinary progress
that has been made over two and one half millenia, particularly in
the last few centuries* `

And then he goes on to say that `*Some fundamentally new insights
are certainly needed* `

What I am trying to say is : If the Universe is conscious, it is a
compelling reason to believe it makes sense. So, I take it that there
is a universal (Cosmic) mind that acts as an interface between the
external world at one end and consciousness at the other end and
in this way ensures permanent consciousness for us all. Here the
word `permanent` is not to be used in the literal sense. Instead, it
is implied that there may be periods of unconsciousness, but the
individual soul is unconscious of these unconscious tenures, so they
pass quickly, rather instantaneously.

The ultimate purpose should be to develop a philosophical model
that is aimed towards creating an understanding of `Reality`, of
`Life and Consciousness`, and of the `Oneness of the mind`. At
broad level the model of philosophy should look at the possibility
of an existential truth that can ultimately bind the universe
together. In my view the initial condition should be that the
universe is obliged to make sense, towards a unity of consciousness,
that ensures bright future for life in general, including life on the
planet Earth.

In his book `WHAT IS LIFE and MIND AND MATTER`
Erwin Schrodinger has discussed the subject of the ` `Unity of
Consciousness` ` or the ` `Oneness of the Mind` at great length,

with a convincing proof of this concept in a thought experiment (Flickering Lights). The subject is covered extensively in my first book ``SIX WORDS`` where the six words are ``WE ALL HAVE THE SAME MIND``

LINE 5 : The Intensity Of The Field At Any Point Is A Measure Of The Information In The Field.

I discuss this subject from two different perspectives:

a) The role of `Mathematics` in creating `Information` via `Physics`

b) The Information In the Field suggests Our universe might be a Simulated Universe.

First the former :

In the words of the Scientist and Philosopher Paul Davies :``*The traditional relationship, between mathemetics, physics, and information, may be expressed symbolically as follows:*

MATHEMETICS → PHYSICS → INFORMATION ``

My view: Indeed `Mathematics` is the President and the company is called `T

he Universe. For more on the subject refer the piece on `GOD IS A MATHEMATICIAN`.

Now for the second perspective:

The Information In the Field suggests Our universe might be a Simulated Universe.

In the words of the scientist Seth Llyod:

`*Information and energy play complimentary roles in the universe, energy makes physical systems to do things. Information tells them*

*what to do. The primary actor in the physical history of the universe
is Information`*

Seth Llyod goes on to say:

`*The universe is a computer, but a quantum computer whose bits
are qubits (quantum bits). Every molecule, atoms, and elementary
particles register bits of information. Every interaction between those
pieces of the universe process that information by altering those bits.
That is, the universe computes. The universe began computing from
its start, and what it computes is itself`* `

One may ask the question ..`But what does the Universe compute
?.`. Seth Llyod gives the answer : `` It computes its own
behaviour. At first the patterns it produces are simple, but as it
processes more and more information, it produces more intricate
and complex patterns, on the physical side giving rise to galaxies,
stars, and planets, while on the human side, producing life,
language, human beings, society and culture. ``

Llyod explains further, that a classical computer cannot simulate
a quantum mechanical system, so the Universe being quantum
mechanical , cannot be considered a classical computer, and can
only be a quantum computer.

My own idea is that all this happens in a field, and when sufficient
information is accumulated in the field, it becomes an `Intelligent
Field`, where the Intelligence in the field keeps increasing all the
time, and this intelligence like that of an `Infinite Mind` is
focussed on creating biochemistries suitable for creating awareness
and consciousnesses etc to understand itself. Hence that portion
of the computation which the universe does, which is attributed to
the intelligence of the Intelligent Field is not completely in the

domain of `quantum computation`, to some extent it could even be called `classical computation`.

It could also be that the Intelligent Field is itself a product of simulation which then simulates the universe?. and it goes on and on, step by step, trillions and trillions of intermediate steps, through trillions of years and countless eons of the universe. And with each simulation the complexities increase. It must have been very very very simple in the very very very beginning. A universe from nothing at all. It could be that we the human beings are an `Intermediate step`.

This viewpoint also encapsulates the Anthropic Principle and explains what happens when a new eon of the universe is created. All the Information is available in the field, all the constants of nature are already known to the Intelligence in the Field, its just the `switching on` that is needed.

Notwithstanding the fact that at the current level of our understanding it is impossible to establish the number of Aeons that preceded the current Aeon of the universe, and the manner in which Simulation took place and under what mechanism could the constants of nature have been implanted in the system that `Switched On` the big bang, the fact remains that the operation was hugely successful in creating an Eternal and everlasting universe (Aeon of) with life and consciousness evolving and flourishing in the galaxies.

All that is required now is to establish how the universe can make sense, and what can be that philosophical model that ensures permanent consciousness for us all.

THE UNIVERSE MUST MAKE SENSE

Simulation or no simulation, we are now in a Universe which can last forever, where life and consciousness is evolving and flourishing in the galaxies, but unlikely to last beyond a certain length of time which is but an insignificant part of the total life of the universe.

All that is required now is to establish how the universe can make sense, and what can be that philosophical model that ensures permanent consciousness for us all.

The laws of science must be such that the `Universe should make sense`, and for the universe to make sense, there must be a consciousness to observe and to understand that the universe makes sense.

But is our Universe obliged to make sense?

In other words, is making sense a requirement that the universe must fulfil?

Somehow it looks as if Scientists are not too much interested in answering this question. And there are some - and that includes some great scientists – who believe that the universe is under no obligation to make sense. According to them, what we know about the universe is perfectly consistent with a multiverse generally inhospitable to life.

As for me, I am NOT a member of this club of scientists.

Scientists will someday have a complete understanding of what is dark matter, what is dark energy. Scientists will someday have a theory that explains everything about matter and energy. But can that theory be called as a `Theory of Everything` can we say that will be the end of physics.

I don`t think so.

Any TOE that does not take into account Mind and consciousness and its impact on Matter and energy is incomplete.

Any theory that does not guarantee a continuity of a universe with life and consciousness is incomplete. In fact this aspect must be set as a precondition for arriving at any meaningful TOE. Any theory that leads towards a future of perpetual nothingness and ends there, is unacceptable to the thinking mind. We must look at hundreds of alternatives that prevent such an end. There is no hurry, there is ample time. The current universe is still in its infancy. Let us keep watching the phenomena. Let us keep flooding ourselves with information.

There are inherent complexities here. Even the greatest Scientists differed on conclusions even though the inputs available to them were similar. Einstein was in perennial disagreement with Neils Bohr on the `Completeness/Incompleteness of Quantum Mechanics` issue. John Wheeler and Wigner were involved in a major argument, on what happens inside a black hole, during the 1979 Centennial symposium to celebrate the achievements of Einstein. Stephen Hawking disagreed with Roger Penrose on Objective reduction of the wave function and its role in the operation of the brain + on the necessity to explain consciousness.

But we can squeeze the complexities and arrive at certain simple statements that do not contradict established truths and are in the domain of possibilities.

In the words of John Wheeler:

``In time to come, a single simple sentence will explain the strangeness of the Universe and as we say that sentence to each other we`ll say-Oh how beautiful- how could we have missed it all that time?` `.

One such candidate for a single simple sentence is this : ` `WE ALL HAVE THE SAME MIND` `

Its just that our consciousness is in the singular.

And so, I come to the first sentence of my presentation:

Imagine a ` `travelling cosmic mind` ` that can ensure a permanent everlasting consciousness for us.

An intelligent Field which is forever and always on the lookout for creating an appropriate biochemistry to give itself consciousness and then understand itself and the universe, and hence allows the universe to make sense.

We ask the question : Why `Travelling`? Is this `Intelligent Field` not Omni present in Time and Space as is the ` `Electromagnetic Field ` and the `Gravitational Field` ?.

To answer the question we need to understand the future of the universe. As per the existing theories of the universe, notwithstanding the fact that they might still be in the domain of philosophy, there is a reasonable consensus of opinion that there are five distinct stages in the life of the universe, in each of which the universe looks completely different.

To begin with the universe has a long life …in excess of 10 ^ 100 years ie more than ten thousand trillion trillion trillion trillion trillion trillion trillion trillion years, and the five distinct eras of the universe are as follows:

The first period is called the `PREMORDIAL ERA` which lasts about three hundred thousand years. No stars, no galaxies.

nothing but radiation, no atoms formed, extremely high temperatures, reducing all the time.

The second period is called the `STELLIFEROUS ERA` which begins as the Primordial Era ends and lasts for about a thousand trillion years ie about

$10 \wedge 15$ years. We are currently living in this era and the current age of the universe is about 13.7 billion years which is less than 1 in 70000 parts of even the Stelliferrous Era. So the Universe is currently just an infant, or maybe a new born baby. What happens in the Stelliferrous Era? Galaxies are getting formed… Stars are shinning, Life and consciousness is flourishing. We are in the Stelliferrous Era. Liquid water is available and carbon based life is possible at many locations (Star systems with planets) in this period.

The third period is called the `DEGENERATE ERA` . Stars stop shining and start dying and the dead stellar remnants such as Brown dwarfs, White Dwarfs, Neutron stars and Black holes constitute the inventory of this era. This period lasts till about $10 \wedge 40$ years (since the big bang). About a hundred trillion years might be a common overlapping period between the two eras. Many of these dead stars, collide with each other, scatter into space, and into nothingness. Of these only The outer atmospheres of large White dwarfs which can support interesting chemical actions, are possible candidates for supporting some sort of an abstract life The energy source could be the radiation field heating from within. Though Carbon, oxygen, and many other elements required for supporting life will be available. Liquid water will be missing, and that is why the nature of life will be quite different from what it is on earth.

The fourth period is called the `BLACKHOLE ERA` which lasts for about trillion trillion trillion trillion trillion trillion years going all the way upto 10^{100} years,

since the big bang. Black holes inherit the universe, warp space and time, evaporate their mass energy, and make an explosive exit.

Life is extremely difficult in this longest lasting of the four Eras, unless life is of a highly abstract nature, with actual matter not taking any part in the proceedings. According to Freeman Dyson the metabolic rate of an abstract creature is proportional to its operating temperature and so is the rate of consciousness. The largest temperature accessible at black hole surfaces is about a billion times smaller than the operating temperature of a human being. Accordingly, the rate of consciousness is slower by a factor of several billion.

The Final era is called the `DARK ERA`.. goes beyond 10^{100} years .. looks like never ending .. nothing going on here except photons moving here and there

We ask the question : For how long can we expect life and consciousness to exist with a reasonably high rate and degree of consciousness ?

A reasonable estimate is 10^{15} years ie upto the end of the Stelliferrous era. However even if we consider an overlapping period lasting about a thousand times longer than the Stelliferrous era into the degenerate era (which is highly improbable) The most optimistic estimate is worked out as 10^{18} years since the big bang.

Now compare this with the least optimistic estimate of the life of the universe at 10^{100} years ie completely ignoring the Dark Era of the deepest future.

What do we get ?

That life and consciousness flourishing with a good rate and degree of consciousness, lasts for only one unit of time out of nearly 10^{82} units of time.

Does this make sense at all ?

And even in this short spell of time, there is nothing very special or unique about the human being. He has to compete with, may be a thousand other intelligent civilizations, many of them may be technologically more advanced than us.

Does that make any sense at all ?, that intelligent life with a reasonably high degree of consciousness exists only for one unit of time out of nearly 10^{82} units of time. Does the universe make sense at all ? ... or can we say that making sense is not a requirement that the universe must fulfill?

But we are conscious, we have always been conscious (even if there were intervening periods of unconsciousness, we were unconscious of these unconscious tenures, so they passed quickly). Indeed, we are forever conscious for we do not remember ever asking the question: `Where is the universe for me?

In short it cannot be that the universe does not make sense.

How does the `Intelligent field` resolve this dilemma?

I can only speculate ... with a wild imagination ... but there is a good chance I may be right ... for that's the only way out. That this mind of ours is nothing other than a ``TRAVELLING COSMIC MIND``

And so I say:

`There is no point in going further into the deep future if I can`t find consciousness there .. let me go back in time again`

To remain forever conscious, the only requirement is that we remain forever in the Stelliferrous Era where Stars are shining and life and consciousness is flourishing. And the only way this can happen is, if we consider our `mind` to be integral with a cosmic mind that can travel backwards and forwards in time and thus remain forever in the Stelliferrous Era. In short if our mind is an omnipresent `Infinite mind`, omnipresent not just in space but also in time.

In short, the mind doesn`t travel into the deep future, for the simple reason that it cannot obtain consciousness over there.

But is that possible?

I should say It is distinctly possible, in fact it is impossible to be impossible.

No doubt of course, that the concept of the traveling cosmic mind is a challenging one; it is very difficult to prove but, I guess, even more difficult to disprove. It has several positives, such as the following:

1) It does not violate any of the established principles of science. On May 31, 2016, I gave a talk on the subject of the traveling cosmic mind at the prestigious National Institute of Advanced Studies in Bangalore; it was attended by several eminent scientists and philosophers of Bangalore. No doubts were raised on the idea during the question-and-answer period; in general, it was well received.

2) It dispenses with the requirement of a multiverse theory, which requires the existence of millions of universes, so that at least one universe has absolutely the precise mathematical constants of nature (as in this universe) so that life and consciousness will appear and then understand the universe. One universe, with all its aeons, is all that is required, in which the cosmic mind travels back and forth in time to remain forever in the Stelliferous Era to obtain consciousness and does not proceed toward the Degenerate, Black Hole, and Dark Eras of the deep future, as there is no consciousness available there.

3) It does not require the big bang to be the time zero (i.e., the beginning) of the universe as an act of God. It assumes the requirement of a long history of several previous aeons of the universe, leading in the end to a primordial consciousness—just ontologically prior to the physical realities of the current aeon of the universe—that contained the coded information for constructing a possible new universe. That coded information constitutes the design of the new universe, complete with all the fine-tuned mathematical constants built into the program.

4) It is in complete agreement with the current thinking (among some of the greatest scientists on the planet) that this is a simulated universe, or a quantum computer. As the scientist Seth Lloyd says, "It computes its own behavior. At first, the patterns it produces are simple, but as it processes more and more information, it produces more intricate and complex patterns, on the physical side giving rise to galaxies, stars, and planets, while on the human side, producing life, language, human beings, society, and culture."

5) It is in complete agreement with the concept of the `Oneness of the Mind`. It's just that our consciousness is in the singular; it's

a traveling cosmic mind that gives us everlasting consciousness. An "intelligent field," which is forever and always on the lookout for creating an appropriate biochemistry to give itself consciousness, thus allows the universe to make sense.

6) It gives a meaning to the universe. Without the traveling cosmic mind, the universe would be completely meaningless; life and consciousness would exist for only one unit of time (the Stelliferous Era) out of nearly 10^{82} units of time, as explained earlier.

7) It is in complete consonance with Einstein's relativity principle (ERP) and adequately takes care of explaining all the various "causality violation paradoxes," which ensue as a consequence of ERP. This aspect is covered extensively and comprehensively in a chapter entitled ``Relativity And The Cosmic Mind`` in my book ``INTELLIGENT FIELD``.

8) It is strong and intelligent and will keep producing islands of negative entropy for lives and consciousness to flourish and understand the universe, not just for a while, but for all times to come. Without it, the arrow of time will always point to the dissolution of structure into a featureless state of maximum entropy. Indeed, the traveling cosmic mind comes to our rescue and saves us from that bleak future.

9) It explains the concept of quantum entanglement perfectly. The mere fact that there is such a thing as quantum entanglement implies instant connectivity, and this connectivity is impossible to imagine except by the consideration of an omnipresent mind (omnipresent in time as well as space).

The topic of my presentation is : Science, Religion, and the Intelligent Field. So far I have spoken about Science and the

Intelligent Field, but not about Religion. Now, let me take a few more minutes of your time to talk about `Religion`, Rather on what is my understanding of what the Essence of religion should be.

Now some people are advising me to be more pragmatic in my approach and not be too excited about the traveling cosmic mind (particularly the traveling part). Some have said that all these ideas will remain unverifiable till our present limited consciousness evolves ultimately into infinite mind (IM). However, all have agreed with the idea that mind and consciousness have a central place in the ultimate nature of reality, never mind whether the said idea is not professionally useful to contemporary scientists, or practically useful to build machines. It can be philosophically useful to unite science with religion, to unite people, to unite cultures, to unite religions, to end conflicts and wars, and so on.

Notwithstanding the power of the traveling cosmic mind in its capacity to give us everlasting consciousness and awareness, the same is powerless, when it's a question of exercising control over the consciousness when it becomes a part of a living organism, even if that living organism is a human body.

Talking about consciousness, in the words of the Scientist and Philosopher Erwin Schrodinger:

`On the one hand it is the stage, and the only stage on which this whole world-process takes place. On the other hand we gather the impression, may be the deceptive impression, that within this world-bustle the conscious mind is tied up with certain very particular organs (brains), which while doubtless the most interesting contraption in animal and plant physiology are yet not unique, not sui generis; for like so many others they serve after all only to* *maintain the lives of their owners, and it is only to this that they owe their having been elaborated in the process of speciation by natural selection`*

But in case of the human mind, as it evolves towards an intelligent human being, in the words of the Cosmologist Ed Harrisson:

`After a span of time natural selection must operate on a different scale altogether. Indeed, we are the outcome of natural selection, and to the operation of this law can be attributed the fitness of the human body. When a civilisation gains control of its environment, the evolutionary game must change, and new rules determined what is fit and unfit. Natural selection must operate in a different way. According to the new law ``Intelligent life forms that are destructively aggressive cannot – or should not - control the environment``. This law must operate conceivably in two modes; the first is unconscious and automatic and the second is conscious and deliberate`

There is every hope that the evolutionary process – with the kind help of the great scientists - will lead towards `Super -intelligent human beings` in the future in total control of the environment. T

Now, there are conflicts between nations which are predominantly Religion based. These can and should be easily resolved by presenting a better understanding of the requirement of a universal religion, which will have no location in place or time, which will be infinite like the God it will preach. It will be a religion which will have no place for persecution or intolerance in its polity, which will recognize divinity in every man and woman, and whose whole force will be created in aiding humanity to realize its own true divine nature. The power of religion, broadened and purified, is going to penetrate every part of human life. Therefore, religions will have to broaden. Religious ideas will have to become universal, vast and infinite.

Adopting a Universal Religion is the prime requirement, and even if there are differences of opinions about the nature of God, or the form in which God exists, all religions must converge into a universal religion that teaches us that God is a mind. I believe that `One ness of the mind ` or the `Unity of consciousness` is the central idea of universal religion.

Whether to call it as `Intelligent Field` or an `Infinite Mind` or `God` is of course just a matter of taste. The Theists can call it `God`, and the Atheists should be happy to call it as an `INTELLIGENT FIELD` or an `INFINITE MIND`. The dispute should end.

Each individual on the planet must have the freedom to choose his or her own religion and that freedom should include freedom from Religion.

So, while there are various deep states attributed to the human beings, Unity of consciousness implies that no individual human being is to be blamed, nor any nation is to be blamed. A nation is just a geographical area. It is the interactions of the world that make people what they are, and how many of them become victims of time. It is the deep state phenomena that needs to be controlled.

In essence every nation is predominantly a `GOOD STATE`, and now is the time for the `GOOD STATE` of nations to first control and then eliminate their `DEEP STATES`. And there is no need for the 9/11 story to be a stigma for the American public, it is best forgotten as an aberration of the past. But the public must be aware of the story and be conscious and aware that the Deep State must not be permitted to stage-manage another war on the planet Earth by Random manipulation of events here and there.

Let us attribute this worldwide deep state phenomenon to the existence of an ``UNINTELLIGENT FIELD`` out there.

So, there is an `Intelligent field`, and an `Unintelligent Field`, and there is a war between the two.

And the `Unintelligent Field` is leading.

So, what do we do?

Transformation is the answer. Transformation of the world leaders who are in need of transformation. Transformation of the World Polity to ensure resolution of conflicts and wars, to prevent Bio Medical Wars, to prevent Trade Wars. Transformation of all religions into a single religion whose essence is:

1. To find solace and peace of mind in prayer and/or meditation.

2. To Live in peaceful coexistence with our fellow human beings

3. To Take care of the environment around us.

4. TO LIVE AND LET LIVE

As I mentioned earlier, while there is an intelligent field in control, the amount of intelligence in that intelligent field is a measure of the information in the field. The current information reveals that the intelligent field is severely contaminated. The present crop of world leaders is unable to deal with the contamination; some are even causing it.

Time is running out for us if we are to save the human race from extinction. The cosmic mind will not come to our rescue, and the beautiful minds of the world must come forward with novel and intelligent ideas that can save us from the perils faced by our species. Leaders of nations must come forward to implement them. Once there was a Manhattan Project, which saved the planet from the very real danger of a fascist world. Now, a new Manhattan Project is required to explore other forms of power and paths to peace.

So, I tell the world:

"Avoidance of extinction must be the most important subject taught in universities worldwide."

THANK YOU

it is my optimistic conclusion that the very personalities whose views and actions are currently pushing humanity closer and ever closer to the brink of war and extinction are also the critical personalities whose transformation is essential for pulling humanity back from the brink to the path of peace, survival and continued progress in our evolution. It is obvious that the current set of leaders of Superpower nations are unlikely to be replaced in the near future. So, how does one limit their potential for harm to the world and its survival? It looks like the only solution that comes to the mind is their transformation.

So, we are looking at two possible scenarios:

1) Good sense will prevail. The world leaders will see the light, come together, and adopt a unified approach to end conflicts and wars.

2) The war continues, gets escalated, turns nuclear, and brings about a nuclear catastrophe.

TAKE YOUR PICK

Today is the 10th of September 2023, and I am very happy to say that good sense appears to be prevailing, that World leaders are seeing the light, they are coming together, and they are making all out efforts to adopt a unified approach to end conflicts and wars. And the country that is showing the light to the world leaders is India.

I am referring to the G20 meeting currently going on in India, under the Presidentship of the Indian Prime Minister Narendra Modi, where India has scripted history, and stamped its Authority on the global stage, securing a consensus on more than 100 issues on the agenda, including the most contentious one, the Russia-Ukraine war, as G20 leaders adopted the New Delhi Leader`s Declaration. The New Delhi Declaration avoided any mention of Russia on the Ukraine war except in the context of efforts to revive the Black Sea Grain Deal.

The Prime Minister Narendra Modi called upon G20 countries to remedy the `Global Trust Deficit`, caused by the pandemic, and the conflict in Ukraine, and said a unified approach alone could help the world devise solutions to the multiple crises facing it. ``TRANSFORM TRUST DEFICIT INTO CONFIDENCE: PM``, was one of several Headlines in Today`s Times Of India, Bangalore edition. The biggest headline in the paper was: `GLOBAL CONSENSUS, MADE IN INDIA`

Western and Japanese sources, however highlighted a new Ukraine-related text that said all states must refrain from the threat or use of force to seek ``territorial acquisition against the territorial integrity and sovereignty or political independence of any state``. While the declaration also said the use or threat of use of nuclear weapons was inadmissible, it again did not name Russia.

Endorsing the joint declaration, Russian G 20 Sherpa Svetlana Lukash said the situation was now reflected in a balanced form and that the ``collective support`` of BRICS countries worked in its favour.

China is learnt to have joined Russia in resisting any attempt to blame Moscow alone for the war.

This leaves Ukraine as the only country that is unhappy with the G20 resolution. ``ALL ENDORSE, BUT NO TALK OF RUSSIA, Nothing to be proud of, Says Ukraine`` is another TOI Headline.

This significant development leads to the conclusion that the Onus to end the Russia- Ukraine conflict now rests squarely on Ukraine.

In short .. it rests squarely on Zelensky.

ZELENSKY NEEDS TO CHANGE

He is currently basking in the glory of being considered a hero for his country. and he is very happy with that, never mind if his country is being destroyed, and his people are dying, or living in fear, or fleeing the country. But this cannot go on. Zelensky will have to choose what kind of hero he wants to be. He can still be the hero who is capable of self-doubt, self- realisation and retrospection, a hero capable of linking his destiny with an ethical conscience capable of realizing that enough is enough, that the war and its continuation has turned out to be catastrophic for the people of Ukraine, capable of realizing that Ukraine is being exploited by the American Deep State, capable of realizing that after all is said and done it was a huge mistake to tilt towards NATO. He must realize that prior to the war in Ukraine about 40% of the population of Ukraine agreed that "Russians and Ukrainians are one people", and now that percentage has dropped to less than 10 %.

There is no doubt he has a tough and challenging job to tackle, considering Ukrainians today are prouder to be Ukrainians than prior to the invasion and are further than before from feeling part of the Russian people. But there is no choice. The challenge has to be surmounted. Perhaps the Russian President Vladimir Putin has a role to play in this. Perhaps he needs to create a congenial atmosphere which will pave the way for reuniting the Russians with the Ukranians – who have the same religion, same culture, same ideology, and are not enemies – and distancing Ukraine from NATO. Perhaps Russia should come forward to repair and rebuild Ukraine to bring it back to its pre-war state.

....

INVENTORY OF SOULS

After all the Information in the Field about the Geopolitics of the planet Earth has been processed ... after all the philosophies have been discussed, let us take stock of the situation and figure out what lies ahead.

Consider the statement:

``Nearly every human being who was ever at any stage alive on this planet earth is alive today``?

All the souls that were once entrapped in human bodies, kept coming back, albeit in new clothes, and perhaps after taking some short breaks in between lives, they are still there.

Let us take an inventory of souls.

Currently there are about 7.5 billion souls entrapped in human bodies, and maybe a few hundred thousand taking a break and preparing to re-enter new bodies, and quite a few new members arriving from different species after passing the qualifying tests.

In the year 1900 the number of souls was just about 2 billion. They have all come back and are among the 7.5 billion, the other 5.5 billion arrived from different life forms.

One hundred years from now if somehow there are no nuclear wars and the population keeps rising on expected lines, the number of souls in live human bodies is expected to be about 18 billion, all 7.5 billion available now will be there, in addition about 10.5 billion are expected to arrive from other life forms. For obvious reasons, this will not be sustained, and nuclear wars will be unavoidable.

Four possible scenarios are discussed ... but there can be many other possible scenarios lying somewhere in between.

SCENARIO ONE:

The most optimistic one where good sense prevails urgently. To be discussed in the end.

SCENARIO TWO:

Now if in the next few years, there is a nuclear war followed by some sort of chain reaction, resulting in the death of about 1.5 billion human beings, we will be left with about 6 billion souls inside human bodies and 1.5 billion of them outside .. hovering somewhere in the atmosphere... and hoping to re-enter back to life soon after taking a short break. But the trouble is ... there will not exist an environment to take back more than just a few of them in the immediate future. Others will have to wait for long time, perhaps several years. The current Intake of souls per day (Number of births) is about 200000. This number will reduce to about 200000*6/7.5 = about 160000 per day, perhaps only to about 140000 per day due to the shocked populace, consequent on the nuclear wars. Rate of intake will of course increase again with time. Still It could take nearly 9000 days ... about 25 years' time before all these souls are taken back. That's a pretty long waiting list. During this period about 1. 5 billion of the available 6 billion souls will come out of the bodies and join those in the waiting list but at the rear end.

SCENARIO THREE:

And what if, near the end of this century.. say a war lasting for five years from 2095 to 2100, there is a major Nuclear war followed by a global chain reaction and only about 1 billion out of about 11 billion survive and 10 billion souls come out into the atmosphere, with nowhere to go, and nothing to do except wait for perhaps several hundred years for a chance to come back inside a human body. It is also possible these one billion survivors may plan for themselves a golden age and enact strict laws to ensure only a gradual increase in population @ say 1% per year and then fix an upper limit of say 4 billion or so. This limit would probably be reached in about 140 years. And there is no guarantee that the original 1 billion will all be among the 4 billion. In all probability a good number will come from the 10 billion stranded earlier on. So it turns out something like this : from the year 2200 onwards 4 billion souls will be inside human bodies and another about 8 billion souls hovering in the atmosphere. Assuming average life expectancy as about one hundred years, it will mean an average waiting period of 275 years.

Still not too bad provided all these souls waiting in the atmosphere do not have an awareness of getting bored and impatient.

SCENARIO FOUR:

And what if ... a few centuries from now ... or even before that ... with about 14 billion as the World population at the time .... There is a surge of Catastrophic global nuclear wars followed by radiation reaching out throughout the globe. All 14 billion die and all the souls are out in the open. No human beings anywhere, no environment or biochemistry available for obtaining human consciousness.

Where will the 14 billion souls go to? Stranded in Interstellar space ... I think ... It will be quite a while ...perhaps tens of thousands of years before a similar biochemistry is available again on the planet Earth. We do not know to what extent or to what degree will the souls be aware of themselves. We know there is an

`Intelligent Field` out there ... for even a single celled creature such as a paramecium is smart enough to swim towards food, negotiate obstacles en-route and try to escape from danger. But the quantum of Intelligence is a measure of the information in the field, and at this stage we cannot say what that information will be available in the field, when we are stranded in space following the self-destruction of the human species. Whether we should wait patiently for that biochemistry to evolve again ... or lose patience and hitchhike to another star system and hope for the best.

Or may be ... get inside Termites ... plenty of food available.

A chilling thought, nevertheless.

What`s more ... The `Cosmic Mind ` will not come to our rescue.

So at any cost, Scenario Four must be avoided ... and the best bet is to remain in Scenario One .. ie GOOD SENSE MUST PREVAIL. An integrated approach must be followed to ensure the population takes a U-turn at the 10 billion level and then stabilizes at the 7.5 billion mark.

An integrate approach is also required to ensure that Nuclear Wars are completed avoided.

BUT TIME IS RUNNING OUT

PRAISE FOR AUTHOR'S BOOKS

SIX WORDS

``Six Words is an autobiographical epic story – epic in the dictionary sense to mean "extending beyond the usual or ordinary, especially in size or scope."

Sagar begins his personal journey at the Big Bang origin of the universe with "I am a quark". He progresses to become an atom of hydrogen, then helium, and finally explodes out of a supernova toward earth as a carbon atom as our planetary system has formed. On earth he becomes an organic molecule and after millions of years he finally becomes a structural engineer.

This book is a tour de force of the major physical sciences, theology, and philosophy. Sagar goes deeply into each, explaining in clear terms very complex subjects. The book then moves to a hypothetical seminar in which the major scientists and philosophers gather to compare notes and thinking. Einstein is there along with Erwin Schrodinger, Charles Sherrington, John Wheeler, Eugene Wigner, and many other top scientists from history.

This is a major five star book written by a serious student, thinker, and observer of the sciences of the universe. The book's aim is to draw religion and science together in a way that leaves established scientific laws and rules intact.

The Six Words do exactly that. I will leave it to the reader to have the joy of uncovering Sagar's Six Words. ``

Dr. Clifton K Meador...Author of the best selling ``Fascinoma``

"...a treatise that ponders the laws of physics, the history of the cosmos, the nature of God and the fate of mankind.

"...kind of "autobiography" of his existence, starting with the formation of his constituent subatomic particles..." "... It begins with a brief, engaging account of cosmology from the Big Bang through the evolution of life. The book then turns to more involved (and less successful) explorations of advanced physics, including the mysteries of Heisenberg's uncertainty principle, "quantum entanglement" and the relativistic paradoxes of travel near the speed of light..."

"... The book's sixth chapter comprises a fanciful "seminar" of great thinkers—from Immanuel Kant to Albert Einstein to contemporary physicist Freeman Dyson..."

"... All this background sets up a section on Sagar's own philosophical speculations, which mix such topics as the anthropic principle—which says that fundamental constants must be able to support the life-forms that observe them—with the quantum mechanics mysticism..."

"...Sagar theorizes that God is an abstract "all intelligent omnipresent...infinite mind"; that humans may eventually merge into the divine "Super consciousness"; and that our main task is to avoid blowing ourselves up in the next few centuries—a disaster that Sagar considers a near-certainty unless everyone works for world peace.``

KIRKUS REVIEWS

``In `SIX WORDS` the author SURENDRA KUMAR SAGAR integrates the works of some of the greatest scientists and thinkers into a discussion in a parallel universe. He fuses quantum physics with psychology to develop a philosophy that brings Science and Religion together ``

THE HINDU

``Words of Wisdom – Science, Philosophy and Religion come together in Surendra Kumar Sagar`s fascinating debut `Six Words`. Sagar uses quantum physics as the backdrop, against which he investigates intricate philosophical questions. Impressive in its scope, Six Words discusses some of the most pressing topics of our time, including the existence of a probabilistic universe, the notion of 'one mind' and the meaning of God and Religion. Sagar ultimately manages to raise the question of how humans can come together to change the course of history."

BANGALORE MIRROR

INTELLIGENT FIELD

The present book `IF` by Sagar is in many ways a continuation of his previous work ``Six Words``. The foundations of ``IF`` may be traced in ```Six Words``. Both the books present an unusual mix of western philosophy, modern physics and religion in a background of the current world polity. Sagar is a keen observer of events in the international arena and the nuclear stockpile seems to be the main concern, which has the capability to annihilate the humans from the face of the earth. When there is a weapon available, a situation for its use will arise eventually; this is the source of concern. Unravelling the inner secrets of a nucleus, man captured the ultimate power which is the source of all energy in the universe. This made him feel invincible. But invincible against whom? It has pitted man against man. History of world is full of wars. Religions have also been used to fight wars and perpetuate conflicts which severely undermine the modern human values. I may however, point out that what we call modern human values are deeply engrained in the great philosophies of Eastern religions since ages. These are not new to us in India. But somewhere in the course of history, we lost them in a struggle for survival. ``IF`` lays bare the underlying struggle which is still on in the world in different forms. Sagar raises important questions about the origins and the hidden nature of these perpetual games of power and survival. Are there any unknown dimensions to these questions and happenings ? Sagar is an engineer by profession. But he displays a deep understanding of the heart of modern quantum mechanics, which he has used to propound the concept of an Intelligent Field. The flow of information in this field keeps everything connected and the evolution of the universe happens in a self-consistent way. Our minds, and the information therein, are also part of this field and therefore, affect each other. This is an interesting concept like a cosmic mind or ``Param Brahma`` in the Vedic literature. Survival of the human race critically depends on the flow of right kind of information in the field. The ``IF``itself may not have any intentions or motives. It cannot, therefore ensure our survival. It is the information that we put in the field and the way we interact, will probably decide the future course of humans. This is where the author tries to make a vehement appeal to all those in the know of things, to attempt to save our extinction. It puts the concepts of mind, consciousness, and soul into focus and compels us to think about the motives of our existence and `If` there is any deeper meaning to it. Keeping the rational thinking and science at the top, it tries to ponder on the ways of merging science with religions of the world. In other words, it argues that what we need is a scientific religious philosophy to save us from the impending nuclear catastrophe. It is a highly readable book with a large number of original references and cross-references to support the ideas presented by the author. It is remarkable that the author has been able to convey his ideas in simple words even though they may appear weird at times. Sagar has persistently pursued these ideas for many decades and the book represents the essence of his thinking and possible solutions to the problems that mankind faces. It is sure to make its mark in the world.

PROF.ASHOK KUMAR JAIN, Award winning Nuclear Physicist and EX Head of the Physics Department I.I.T.Roorkee

----

BRIGHT LIGHT IN THE SKY

` `Surendra Kumar Sagar's third book, "Bright Light In The Sky" follows the trajectory of his earlier books, in being profound, provocative and probing. Reading Mr. Sagar's latest contribution, I am reminded of Edward Gibbon's "Decline and Fall of the Roman Empire" which is a narrative through a thousand years. We see the greatness of that Empire at its height, its military organization, its provincial administration, its welter of races, the rise and clash of two religions, the passage of Greek philosophy into Christian theology. But throughout this 'history' it is Gibbon who speaks. Mr. Sagar's saga resonates with similar vibrations. The ideas, events and instances, described in his book, is not his creation, but the voice you hear is the inimitable Surendra Kumar Sagar, weaving a tapestry of thought and philosophy, ushering a grand confluence of ideas and conundrums, displaying the need for convergence of imagination, as a compass, for our elusive quest for a sense of the future, on the journey yet to be.``

Shoumen Palit Austin Datta .. Director MIT ..Academician, Author, Research Scientist at MIT

``I am very happy to learn about your new book. It is a very interesting one. I wish you continued success in achieving a nuclear peril free world.``

Prof. M.S. Swaminathan ..Founder Chairman and Chief Mentor UNESCO Chair in Eco Technology .. M S Swaminathan Research Foundation

----

``This is the third book of Surendra Kumar Sagar, a trilogy on Nuclear proliferation, possible nuclear war scenarios, International geo-politic and power games, and where the world stands now, all set into a philosophic and scientific background. Mr. Sagar is a structural engineer yet displays a remarkable understanding of the heart of quantum world, its implications and possible connections with the human mind and consciousness, all via the `Information in the field``

Dr ASHOK KUMAR JAIN .. National Award winning Nuclear Scientist

----

``Mr. Sagar`s third book is a true reflection of his own erudite and eclectic self. Since he is a structural engineer himself, his book presents a cornucopia of fabulously structured thoughts and concepts.

But this is not the structure of a typical engineer with linearity or even multilinearity with right angles and rigid geometric shapes. He presents an intricate, complex and aesthetic organic arabesque of a multilevel forest – with sinuous vines intertwined with massive trunks of giant trees and the shrubs vying for space with ground-level gossamers.

Taking a leap of faith, one could bring in several similes and similarities with our own ancient scriptures. The triumvirate of Body, Mind, and Soul is presented in the form of Reality, Information and the Right Value System – GOOD SENSE. The psychosomatic interactions are parallaled with the mutual influencing of Reality and Information and the serpentine DNA-like shape of such influences.

And then Mr. Sagar presents the ultimate conundrum – the very survival of Homo Sapiens a la Yuval Harari. Through a willowy wizardry of equations, it is predicted that if we the humans can manage to survive any nuclear holocaust for the next millennia then we are pretty much assured of a perpetual existence.

In his book, Mr. Sagar plumbs the depths of unfathomable profundity and covers an extraordinary canvas of eternal expanses of Time and Space. Weaving through these unthinkable discontinuities is his very unique concept of the Travelling Cosmic

Mind – which is like the unifying force envisaged by quantum physicists and atomic scientists.

It would be extremely interesting to look forward to Mr. Sagar's fourth book after – SIX WORDS, INTELLIGENT FIELD and this BRIGHT LIGHT IN THE SKY – and wallow once again in a continuum which is simultaneously both intuitive and counter-intuitive !!!

Prof. Shashi Sharma ..Visiting Faculty + Trainer + Advisor for ETHICS and CSR;

SWITCHED ON

` `This is the book we have been waiting for … It is a significant attempt towards developing a better understanding of the Deep State of America, and that of many other countries. These Deep States are able to stage-manage nearly every war on the planet Earth in a way that ensures business as usual for their vast military industrial complexes. Together, these Deep States are all set to create war fronts at several locations on the planet, never mind if the common people of these warring countries do not want war. The author makes a valiant attempt to awaken the conscience and common sense of the human beings so that we can `dismantle` these Deep States of the World and reach a new stage in our lives where we can look back on war as an incomprehensible aberration of our past` `

Prof. M.S.Swaminathan ..Founder Chairman and Chief Mentor UNESCO Chair in Eco Technology. M.S.Swaminathan Research Foundation. Ex President Pugwash Conferences on Science and World Affairs.

DEEP STATE (The Mysterious State Withn The Unided States)

``At the outset, I wish to mention that Sagar`s latest book unlike his previous books (stressing on the scientific aspects of cosmology and the place of mind in the universe) concentrates on the geopolitics of "Deep States" and "Military Industrial complex (MIC)" in the US in particular and in other nations in general. The first half of the full text explains how the Deep States and MIC work in parallel to national governments to thwart all efforts by all concerned to bring about peaceful resolution of all intra- and international civil, political, social, economic, ethno-religious - ecological and other conflicts. What is more, they indeed acerbate such conflicts to such a pitch as to increase the danger of nuclear warfare or ecological Armageddon in the near future to near certainty. In particular, the Deep States and MIC of the US, Russia and China enhance the risk of nuclear war far more than those of other nations. This endangers the continued survival and evolution of mind alone not only alone on earth but prevents the opportunity near at hand for mind on earth to integrate with and ride piggy-back on a universe-wide mind possibly in existence at a higher stage of evolution than ours.

The second half of the text tries to find solution to the problems created by the Deep States and MIC so that the mind on earth is freed to evolve along its path in a smooth and uncluttered manner. Sagar`s bringing in the whole text in the style of Greek dialogues between "revised versions" or reincarnations of path-breaking scientists and philosophers of the past tries to add spice to the issues under discussion without in any way allowing the seriousness of the discussion to sag in any way. The conclusions though fuzzy cannot but be otherwise in an attempt to read and if possible alter the future of humanity and the universal mind towards its best glory. If this book during its publication could reach leaders of all the nuclear nations and nations with vast MICs and the representatives of all other member nations of the UNGA, one hopes there is a chance, however slender, that the earth-wide mind is set on its evolutionary path towards uncluttered improvement to its fullest potential. This chance singularly justifies all his painstaking efforts in bringing out the book. I wish all success to the book while hoping it has the widest possible reach among those who are concerned about peace and progress on earth as also among those who wittingly or unwittingly are threats to such peace and progress so that both their minds are reoriented to merge and move together in unison from here onwards. ` `

A.K.Chandrashekhar .. Retired Finance Executive with interest in Science, Philosophy, and social well being.

IS THIS OUR FINAL MILLENNIUM .. SECOND EDITION

The book is about an imaginary seminar where the world situation is discussed, and major issues resolved. The main thrust of the book is current geo-politics within the broad theme of human survival or extinction on earth common to all the author`s past books. It is an extensive update to cover major events that have occurred or continuing to occur at global geo-political level and their fallout on the chances of human survival or extinction during the current millennium. This canvas has been displayed in the format of imaginary conversations between real persons alive or dead whose views or actions have intimate relevance to the broad theme and the special thrust of the book under review.

What is of particular interest to readers of the book is the author's optimistic conclusion that the very personalities whose views and actions are currently pushing humanity closer and ever closer to the brink of war and extinction are also the critical personalities whose transformation is essential for pulling humanity back from the brink to the path of peace, survival and continued progress in our evolution. What could hopefully achieve this critical transformation is bringing all the critical personalities involved in the major geopolitical, religious and commercial conflicts to the table of dialogue with open mind to end war, shed personal ego and nationalism and solve all the ongoing conflicts with the single criterion of peace and just and best interests of all people everywhere and forget and forgive all the rest that is divisive. Who should be assigned this onerous task of persuading the transformation of all critical personalities? The author gives the answer - men and women reputed for their pursuit of peace, justice and continued all round progress of humanity.

A.K. Chandrashekhar

....

CONFERENCE IN A PARALLEL UNIVERSE

(Not published yet)

In the 27 pages of your foreword, synopsis and epilogues, you have practically summarised the significant aims/contents/conclusions of all your past books with the common theme of saving humanity from itself and its self-destructive tendencies. This latest book of yours gives the justification for all the books in a way that should persuade the opponents, convince the sceptics, and strengthen the views of the supporters of world peace as the most pressing need for preventing consciousness from its demise and allowing it to continue to evolve in the only known part of the universe.

While the series of the author's books on the subject try to inform "the irrational out of irrationality", the nature and ability of the Deep State and the many fused states in the world in its far more determined ongoing efforts to inform the rational out of rationality does appear formidable for the attempt of the author to succeed in its aims. This is particularly so because of the fact that in the democratic West the self-interest of the corporate world is similar to the self-interest of the dictators and oligarchs of the non-democratic parts of the world including Russia, China and their satraps in other nations besides semi-hereditary leaders of Africa, Asia and other parts of the world cannot be changed through persuasion and appeal to reason, public interest or world peace. The indifference of a large section of the intelligentsia, the unknowing masses, and the short-term interest of a vast middle-class may also act as a challenge to such efforts at world peace. As of now even a large majority of the Nobel Peace Laureates and the members of their selection panels have preferred to be silent on the ongoing flames of war of varying complexions in different parts of the world.

In spite of the daunting challenges to the author's efforts, all right-thinking persons would no doubt agree with me that for facing the challenges to world peace there are few options, if at all, but to adopting the aims, optimism and determination of the author. For this and this alone, the author deserves all support and admiration from the votaries of world peace.

Chandrashekhar A K

… … …

I am quite amazed that you have been able to complete this book despite your ill health. I congratulate you and wish you all the best. Your persistent efforts would be quite supportive of the world peace, which is so unlikely to emerge any soon. But the consistency and perseverance shown by you is remarkable.

Ashok Kumar Jain

... ...

ABOUT THE AUTHOR

SURENDRA KUMAR SAGAR was born in 1941 and completed his graduation in Mathematics, Physics, and Statistics from Bombay University in 1959, and in 1964 he graduated in Civil Engineering from Thapar Institute of Engineering and Technology, Patiala. He has about 59 years' experience in the design and construction of a vast variety of buildings and structures both in Industrial, and Real Estate sectors. Currently since 1999, he is Technical Director (Design and Engineering), with Total Environment Group of companies, Bangalore.

Surendra Kumar Sagar has a deep interest in Science, Philosophy, and Nuclear Geopolitics including the philosophical impacts of Quantum Physics and Cosmology. In 2003, he attended a weeklong symposium on `Science and Beyond` held at the Institute of Advanced studies, Bangalore, and met and had discussions with some of the greatest Scientists and philosophers of the World, who were present. He is the author of several books including ``SIX WORDS`` published in 2014, and ``INTELLIGENT FIELD`` published in 2017. In the former, he has tried to integrate the works of some of the greatest Scientists and thinkers – past and present – into a cohesive progression, and by fusing Quantum physics with the psychology of the mind, has endeavored to develop a Standard model of philosophy, which he thinks stands a good chance of fusing Science with religion. In April 2014 he delivered an institute lecture at the Indian Institute of Technology, Roorkee on the subject `Standard Model of Philosophy`. In `Intelligent Field` the ideas of an `Intelligent Field` and of a `Travelling Cosmic

Mind` (One of the Central Themes of the book) as described in the book, have added a significant new dimension to the quest for knowledge on Philosophy, Religion, Spirituality, Humanism, and other similar paths towards formulating an integrative view of life and the universe. In May 2016, he gave a talk on the `Travelling Cosmic Mind` at the prestigious National Institute of Advanced Studies Bangalore, and in November 2017 he gave another talk at the I.I.T. Roorkee, this time on `Intelligent Field`. Prof. M.S.Swaminathan - the architect of India`s Green Revolution, and EX President of the `International Pugwash Conferences`- who wrote the Foreword for `Intelligent Field` considered the book as extremely significant.

Sagar has a penchant for providing lengthy reviews on `Amazon` on books that he enjoyed reading immensely. Some of the books he has reviewed include:

1) `The Road to Reality` by Roger Penrose

2) `Information and the Nature of Reality` edited by Paul Davies and Neils Henrick Gregerson. It is possible his review on this book could be a record longest review on `Amazon`.

3) ` The New Quantum Age` By Andrew Whittaker.

4) `Keepers of the Nuclear Conscience` By Andrew Brown.

5) ` The Critique of Pure Reason` By Immanuel Kant.

Sagar has a deep interest in a wide range of subjects and has written extensively on these in his website ``www.sixwords.in``.

Some examples given below:

1) General Expression for Determination of Waring Limits. (On Theory of Numbers)

2) Sensex and the Y Factor. (On Stocks and Shares and the parameters used in the Analysis to predict the future)

3) Laws of Causation and the `Code Script` - Impact on Horses (On Horse Breeding of Thoroughbred race horses)

4) When Two Worlds Collide (On the likely collision 4 billion years hence with the Andromeda Galaxy)

5) Inventory of Souls (On Philosophy and on what might happen to souls in large scale deaths due to Nuclear Wars)

And many others.

LONG INTERACTIONS IN RESPONSE TO BLOGS ON HUFFINGTON POST:

1) Dark Energy Explained (Response and In- depth discussion with the Scientist and author `Louis Del Monte`)

2) Can Black Holes Tell Us Something About Digital Computers? (Response to this blog by the Scientist Mario Livio leading to a lengthy discussion on the subject `Simulated Universe`.

3) The Disappointing Higgs And Sterile Neutrinos (Response to a blog by the Scientist Victor Stenger)

4) Can the Truth Come Back With a Capital `T` (Response to a Deepak Chopra Blog - on Philosophy)

And many others

Sagar currently lives with his wife Bharati Sagar – a well-known artist - in Bangalore, India. They have two sons and three grandchildren.

…… ……

NOTES

NOTES

NOTES

NOTES

www.ingramcontent.com/pod-product-compliance
Lightning Source LLC
LaVergne TN
LVHW052255210726

843527LV00041B/574